A First Course in TOPOLOGY

An Introduction to Mathematical Thinking

ROBERT A. CONOVER

DOVER PUBLICATIONS, INC.
MINEOLA, NEW YORK

Bibliographical Note

This Dover edition, first published in 2014, is an unabridged republication of
the work originally published by The Williams & Wilkins Company, Baltimore,
Maryland, in 1975.

Library of Congress Cataloging-in-Publication Data

Conover, Robert A. (Robert Allyn), 1942–
   A first course in topology : an introduction to mathematical thinking /
Robert A. Conover.—Dover edition.
      p. cm.
   Originally published: Baltimore, Maryland : Williams & Wilkins Company,
1975.
   Includes bibliographical references and index.
   ISBN-13: 978-0-486-78001-6
   ISBN-10: 0-486-78001-5
    1. Topology. I. Title.

QA611.C683 2014
514—dc23

                                                        2013044347

Manufactured in the United States by Courier Corporation
78001503   2015
www.doverpublications.com

DEDICATION

To A. H. S. and L. D. E. S., without whom this book would not have been possible.

Preface

This book is designed primarily to serve as a basis for a course in topology in which the students prove all or most of the theorems. As such, it can also serve as the basis for a lecture course in which the instructor presents most of the results himself. Probably the best way to use it is for a course conducted somewhere between these two extremes.

It is not just an outline. There is much discussion, as in a regular textbook, and a good deal of this discussion is about how to prove the theorems and why they are important. In addition, some of the harder theorems and examples are given rather complete outlines. For example, I do not think that it is reasonable to expect an undergraduate student to come up with a proof of Urysohn's lemma or the Tychonoff product theorem, nor do I expect that he would discover a (Hausdoff) completely regular space that is not normal—and justify it—without some help. I have found when teaching topology that a very effective method of getting students to understand difficult proofs (and, incidentally, to see that not every proof can be dashed off in a few lines once the method is discovered) is to give them the proof outright from a standard text, and to have them present it completely and in detail in class. The outlines given here are an attempt to improve upon this method.

It is my hope that when this book is used as a basis for a course in which the students present most of the proofs, it will alleviate some of the problems that beginning students have with this way of teaching. For one thing, reading can be assigned and discussed. The student is used to having something to read, and suddenly having nothing at all can be quite a blow. Besides, it is unrealistic to force him to work in a vacuum. Professional mathematicians don't; neither should the student. The problem up to now has been that anything that the student might read at the beginning level contained too much—any supplementary text turned out to be an answer book. This book solves that problem.

Probably the biggest complaint about teaching from an outline concerns content: it is said that the better students learn something about how to do mathematics, but they do not learn much topology, and the average student learns just about nothing. I do not agree with this complaint—not if the course is handled properly—and this book is an attempt to give the beginning student (who, after all, has not had much experience in doing mathematics on his own), something to hang onto until he develops some confidence. Additionally, by relating new concepts to old ones and by spiraling back to topics already covered, this book should help everyone come out of the course with a pretty good foundation in topology, as well as with some experience in doing mathematics.

The question of exactly what a good foundation in topology is seems to be an open one. It is my impression that many students come out of their first course in topology totally missing the point, probably because most books try to be totally rigorous at all times, and will not allow any hand-waving, even at the beginning, thus rendering themselves pretty trivial and pretty static. Topology is anything but static, and most students have some idea of "rubber sheet geometry," but they usually do not see any in most first-year courses. Part of Chapter 3 is an attempt to get the student to think topologically in terms of pictures, as most mathematicians do when they do topology. Students should have a lot of fun with it, and hopefully, they will become excited and curious enough to want to see more. In order to be able to present this "advanced" material so early in the course, we are sometimes forced to resort to descriptions instead of definitions, and some exercises begin with "convince yourself" instead of "prove." The difference is clearly pointed out, and the student is never asked to prove anything until a proof is possible. (There is a recent calculus book which asks the reader to "prove" that a certain function is continuous, without ever giving a definition of continuity.) Every theorem in this book is indeed a theorem which can be proved.

About content. Most of the material is standard point-set topology (except, of course, for the last chapter) which, as Kelley says, is "what every young analyst should know." Continuity is emphasized throughout. The space of countable ordinals is an important example in many cases, an example which is realistic and not "made up" just to illustrate a particular point. However, not everyone will agree that undergraduate students can understand this space at least as well as they understand the space of real numbers, and those who do not can eliminate any discussion of ordinal numbers and still use this book. Examples other than ordinal spaces are presented when it seems feasible, and it is always clearly stated when a particular section, theorem, or example requires using the space of countable ordinals. (Larger ordinals are not used at all.) Similarly, any discussion of product spaces with infinitely many factors can be omitted (with the resulting loss of content, of course); as with ordinals, places where infinite products are used or needed are clearly pointed out. The student is constantly asked to establish examples of spaces that have the various properties discussed, as well as examples of spaces that do not have the property. Before doing so, he is always asked to "state precisely what it means" when a space does not have the property under discussion, or when the particular property at hand does not hold. This serves two purposes: first, he knows precisely what to look for in a counterexample, and, second, he learns to work with mathematical phraseology and to appreciate the precision of a mathematical statement.

There is enough material here for a full year course. However, by picking topics, one can construct a one-semester course as well. With the thought of eliminating certain sections in mind, no topic is introduced solely in the exercises if it is to be used later. Rather, if a topic is needed that was previously discussed in an exercise, it is defined again, the relevant properties are pointed out again, and a reference to its initial discussion is given so that the student can see how it arises naturally.

A student who has had, or is taking, a course in advanced calculus and who is willing to work should be able to handle the material presented here, except for the last chapter, where a knowledge of elementary group theory is required. It has been my experience that most students who take a course like this one become very interested and are willing to work very hard indeed. As a result, they learn a great deal about mathematics, and quite a bit of basic topology as well.

Introduction

Topology is sometimes called "rubber sheet geometry" because two objects are said to be topologically the same if one can be stretched, shrunk, bent, or twisted to make it look like the other (without overlapping itself or ending up being ripped apart). For example, a triangle, a square, and a circle are all topologically the same:

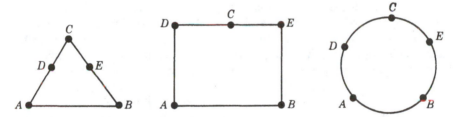

And all three figures are topologically the same as a closed squiggle:

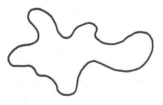

But none of these four figures is the same as a figure 8 or a straight line segment. (Why not?)

Probably the most famous example in armchair topology is that a doughnut and a coffee cup are topologically indistinguishable:

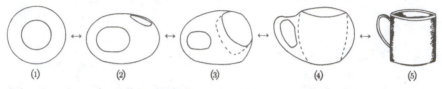

(The doughnut is solid; the indentation that we made in Step 2 is not cut out but instead is "punched" into it as if it were made of clay. The original hole in the doughnut becomes the hole in the handle of the cup.)

Of course there is a lot more to topology than just turning lead into gold. It is a fascinating and exciting subject and a very powerful one: some relatively difficult theorems from calculus, for example, fall out as simple corollaries of relatively simple topological theorems (the "intermediate value theorem," for example, and the fact that a continuous function defined on a closed interval attains both its maximum and its minimum). But it is more than a tool to prove theorems that we already know; topology is one of the most interesting and important fields of modern mathematics.

This book is a brief introduction to topology. Also, it is probably unlike any mathematics book that you have used before because it expects you to *do* mathematics instead of just reading it. Very few theorems are proved in the book (although hints are given in many cases); you are supposed to supply the proofs, even though they are usually not asked for.

Creating proofs and doing mathematics in general can be exciting and it can be fun, but it can also be frustrating. This book attempts to maximize your enjoyment of mathematics and to develop your skill in doing it, but it also necessarily maximizes the frustration involved in learning the subject. Learning is not easy. It has been said that there is nothing more frustrating than having a blank piece of paper in front of you and being expected to fill it with mathematics. You will find that this is true. But it is also true that there are very few things as satisfying as seeing the paper fill up with a good idea that is your own. As you work through this book, you will undoubtedly experience both feelings. If, at the end, the satisfaction outweighs the frustration, then the book will have done its job.

A word about content is in order. The treatment of sets, functions, and transfinite numbers in Chapters 1 and 2 is similar to that given to the real line in most first-year calculus books. It is designed to give you a working knowledge of the concepts involved; it is not a complete development of these concepts. If you get interested and find that you want to pursue any of these subjects more deeply, the book *Introduction to the Foundations of Mathematics* by Raymond L. Wilder ([18] in the bibliography) is an excellent reference. However, you should not look in any other books for supplementary material about any topic presented here until you have completely covered what we do with it. Looking up the proofs is not the way to get the most benefit out of a book like this one.

Proofs in Mathematics

A NOTE TO THE READER

Despite what many students in first-year calculus think, there is nothing mysterious about the proof of a mathematical theorem. The proof simply shows how the conclusion of the theorem follows logically from the hypothesis. It convinces anyone who reads and understands it that the theorem is true.

There is a big difference, though, between reading a proof and writing one; it is almost as big a difference at times as there is between hearing a symphony and being Beethoven. (Not always and not often, but sometimes.) Since the heart of this course is proving theorems, we discuss proofs a little before we start.

First of all, a proof should be written in good, clear language, using complete sentences. It is not good style to write a proof in two columns, one for statements and one for reasons. Of course a reason should be given for anything that is not obvious, but it should be integrated into the proof with words like "since" and "because," and not be set off to the side somewhere. Also, you should not have to apologize for anything that you say: if you have to say "but what I really mean is" then you should say what you mean in the first place. A proof should be able to stand alone, and should not need its author around to explain it.

Of course, even the most elegant style of writing will not save a proof if the reasoning is wrong, and you never want to make a statement in a proof that you are not sure of. It is not a proof until you are *sure* that it is a proof. Above all, you should *never* try to pull the wool over anyone's eyes.

As a simple (probably deceptively simple) example, suppose that we want to prove the following theorem.

Theorem. If all men are mortal and Socrates is a man, then Socrates is mortal.

A good proof should look something like: Since Socrates is a man and all men are mortal, it follows that Socrates is mortal.

You may protest that it does not say anything (and indeed it does not say much because the example is so simple), and you may be tempted to pad it a little: Since Socrates is a man and all men are mortal, it follows that Socrates is mortal because if he were not then he would not be a man. It isn't wrong, but the padding is obvious and should probably be avoided. One kind of padding that should absolutely be avoided is anything like: Since Socrates is a man and all men are mortal and a pound of lead weighs the same as a pound of feathers, it follows that Socrates is mortal. You should never include irrelevant statements in a proof even if they are true. (It is not always so blantantly obvious, and you may find yourself doing it if you are not careful.)

Proving a theorem sometimes involves a good deal of creativity and clever thinking; it *always* involves a complete understanding of what the theorem says and of what you have to work with. Once it is proved, a theorem is another tool that may help in proving a later one, so you should learn and understand each one, even if you are not the first one in the class to prove it.

Topology is a very visual subject and you will find that ideas of how to start a proof will often occur to you if you draw a picture of the situation at hand. It is true that a picture is not a proof, but a well drawn picture is often an indispensable aid to proving theorems in topology; you should draw a lot of them as you work through this book.

Many of the theorems that you are asked to prove are "if and only if" theorems. An if and only if statement is really two statements in one, the "if" part and the "only if" part. For example, consider the statement $2x = 4$ if and only if $x = 2$. The two statements in this single if and only if statement are the if part: if $x = 2$, then $2x = 4$, and the only if part: if $2x = 4$, then $x = 2$. It is not really very important to know which is which; however, hints are often given in the text for one part or the other, and it is easier to use the hint if you know to which part it applies. To decide which is which, eliminate the words "only if" to get the if part, and eliminate the word "if" to get the only if part. Thus, in "$2x = 4$ if and only if $x = 2$," the if part is "$2x = 4$ if $x = 2$" which, in better order, is "if $x = 2$, then $2x = 4$." The only if part is "$2x = 4$ only if $x = 2$." Changing "only if" to "implies," this is "$2x = 4$ implies $x = 2$," which is the same as "if $2x = 4$, then $x = 2$."

Another expression is sometimes used for if and only if. We can say "$2x = 4$ if and only if $x = 2$" by saying "a necessary and sufficient condition that $2x = 4$ is that $x = 2$." The necessity is the only if part, and the sufficiency is the if part.

Some people abbreviate "if and only if" by "iff" or by "⇔."

One final word about the hints. There is almost always more than one way to prove a theorem, and you may come up with a different proof than the hint suggests. By all means, do so! The hints are meant to help you get started if you need some help, and are definitely not meant to limit your imagination. Besides, your proof might even be better than the one suggested.

Contents

chapter one

Sets and Functions

You have certainly had much experience by now with both of the concepts in the title of this chapter. But it is a fact that a thorough familiarity with sets and functions is essential to any understanding of modern mathematics, so the material in this chapter is of basic importance to all of your future work. You should read it carefully, do all of the exercises, and prove all of the theorems.

1. SETS

You may recall that, in Euclidean geometry, the words "point" and "line" are not defined. Instead, it is assumed that everyone has a good idea of what points and lines are, and we let it go at that. This failure to define terms is not because of laziness or sloppiness, though; the fact is that in any system of human thought we have to start somewhere, so there must always be some primitive notions that cannot be defined because we have nothing to define them in terms of.

In this course, the most basic notions that we will use are those of set, element of a set, and what it means for a given element to belong to a given set: we will not define any of these ideas. But of course it is all clear: a set is a collection of objects, an element of the set is one of these objects, and it is always clear when a given object is an element of a given set. For example, the set of all people who are now citizens of the United States is the collection of all those people, and only those people, who are now citizens of the United States. The set is the collection of all U. S. citizens, an element of the set is a U. S. citizen, and one can always determine whether a given object belongs to the set: the given object must be a person and that person must be a U. S. citizen.

Two sets are the same if and only if they have exactly the same elements. Formally,

1.1. Definition. Let A and B be sets. Then
1) A is a subset of B, written $A \subset B$, if every element of A is also an element of B.
2) A and B are equal, written $A = B$, if both $A \subset B$ and $B \subset A$.

Notice that $A \subset B$ does not imply that A and B cannot be equal. If $A \subset B$ and A is not equal to B (in which case we call A a **proper** subset of B), we sometimes write $A \subsetneq B$, for emphasis.

A word about definitions is in order. A definition in mathematics is always an "if and only if" statement. Thus, looking at Definition 1.1, for example, if we know that both $A \subset B$ and $B \subset A$, we can immediately deduce that $A = B$; conversely, if we know that $A = B$, we also know that both $A \subset B$ and $B \subset A$. The most important thing to us about a definition is that it is to be *used*. For example, to prove that two sets are equal, all we have to do is show that the definition is satisfied (i.e., that each is a subset of the other). But—and this is where trouble often occurs—satisfying the definition is not only all we *have* to do; it is what we *must* do at this stage in the development of the subject. Later on we will have theorems that we can use as well as definitions, but for now there is no magic way to avoid the definition, and no reason to want to.

For the moment, then, to show that two sets A and B are equal, the *only* thing that will work is to take an arbitrary element from the set A and show that the properties of A and B imply that this element also belongs to B; this shows that $A \subset B$. Then reverse the process: to get $B \subset A$, take an element from B and show that it also belongs to A. The two containments ($A \subset B$ and $B \subset A$) then allow you to conclude that $A = B$, by invoking the definition of equality for sets.

1.2. Exercise.
    State precisely what it means when two sets A and B are not equal.

2. NOTATION

When an element a belongs to the set A, we write $a \in A$, and when a does not belong to A, we write $a \notin A$. If $P(x)$ denotes a property that the object x may or may not have, then we use

$$\{x: P(x)\}$$

to denote the set of *all* objects x with the property $P(x)$. Sometimes this is modified to

$$\{x \in S : P(x)\}$$

which means the set of *all* elements in the set S which have property $P(x)$, where S is some given set. Thus, for example, the set $V = \{a, e, i, o, u\}$ consisting of the letters in the English alphabet that are always vowels could be written as

$$V = \{x \in A : x \text{ is always a vowel}\},$$

where A denotes the English alphabet.

An important subtlety that needs to be emphasized again is that $\{x \in S : P(x)\}$ *always* means the set of *all* elements of S with the property $P(x)$.

We will use the ordinary symbols for open and closed intervals on the real line (the symbol **R** is reserved to represent the real line, the set of real numbers). Thus the **open interval** on the line whose end points are the real numbers a and b is

$$(a, b) = \{x \in \mathbf{R} : a < x < b\},$$

and is the set of *all* real numbers greater than a and less than b. The **closed interval** with end points a and b is

$$[a, b] = \{x \in \mathbf{R} : a \leq x \leq b\},$$

and there are two kinds of half-open intervals:

$$[a, b) = \{x \in \mathbf{R} : a \leq x < b\}, \text{ and}$$

$$(a, b] = \{x \in \mathbf{R} : a < x \leq b\}.$$

We also have unbounded intervals, called **rays:**

$$(a, \infty) = \{x \in \mathbf{R} : x > a\}$$

$$[a, \infty) = \{x \in \mathbf{R} : x \geq a\}.$$

Similarly, we define $(-\infty, b)$ and $(-\infty, b]$ (note that with this notation, $\mathbf{R} = (-\infty, \infty)$).

Some familiar and important sets of numbers are used so often that we reserve special symbols for them. These are:

$\mathbf{Z}^+ = \{1, 2, 3, \cdots\}$, the set of **positive integers,**
$\mathbf{Z} = \{\cdots, -2, -1, 0, 1, 2, \cdots\}$, the set of **integers,**
$\mathbf{Q} = \{p/q : p, q \in \mathbf{Z} \text{ and } q \neq 0\}$, the set of **rational numbers,**
    and of course,
$\mathbf{R} = $ the set of real numbers.

In addition, the **empty set** will be denoted by the symbol θ (this symbol is *not* the Greek letter ϕ; rather, it is a zero with a slash). The empty set has an important property that no other set has: it is a subset of *every* set. Proving this is tricky. Try it by contradiction unless you are familiar with formal logic and the fact that $P \Rightarrow Q$ is a true statement whenever P is false. No matter how you try to prove it, you are going to *have* to use Definition 1.1.

2.1. Theorem. For any set A, $\emptyset \subset A$.

Finally, the set of all subsets of a set X is called the **power set** of X, and is denoted by $\mathcal{P}(X)$. The power set is an important example of something that may be unfamiliar: a set whose elements are also sets. Such sets are of fundamental importance in topology and we will deal with them often. As a simple example, let $X = \{1, 2\}$. Then

$$\mathcal{P}(X) = \{\emptyset, \{1\}, \{2\}, \{1, 2\}\}.$$

Note that $\emptyset$ is in $\mathcal{P}(X)$ as it should be according to Theorem 2.1. Also, the usual notation applies even though the elements of $\mathcal{P}(X)$ are sets, and we write $\{1\} \in \mathcal{P}(\{1, 2\})$, for example.

2.2. Exercises.
1) What is $\mathcal{P}(\{1, 2, 3\})$?
2) How many elements do you think that $\mathcal{P}(\{1, 2, 3, 4\})$ has? $\mathcal{P}(\{1, 2, \cdots, n\})$, where n is a positive integer? Why is $\mathcal{P}(X)$ called the *power* set of X?
3) Are the following statements true or false:
   a) $x \in X$ if and only if $\{x\} \in \mathcal{P}(X)$.
   b) $\{x\} \in \mathcal{P}(X)$ if and only if $\{x\} \subset X$.
   c) $\{x\} \subset \mathcal{P}(X)$ if and only if $x \subset X$.

3. OPERATIONS ON SETS

You are probably familiar with the idea of a collection of objects "indexed" by the positive integers, like $\{A_1, A_2, A_3, \cdots\}$. Similarly, we can use other sets as indexing sets. Thus if Λ is a set, then $\{A_\lambda : \lambda \in \Lambda\}$ is a collection of objects (the A_λ's) indexed by the indexing set Λ, which means that there is one element in the collection corresponding to each element of Λ. (With this notation, we could write $\{A_1, A_2, A_3, \cdots\}$ as $\{A_n : n \in \mathbf{Z}^+\}$.) For example, the set

$$\{\tfrac{1}{2}, \tfrac{1}{4}, \tfrac{1}{8}, \cdots\}$$

could be written in indexed form as

$$\{\tfrac{1}{2}^n : n \in \mathbf{Z}^+\},$$

and the set of odd positive integers between 5 and 19 could be written in indexed form as

$$\{2n + 1 : n \in \mathbf{Z}^+, 2 \leq n \leq 9\}.$$

Sometimes the *only* way that a set can be written symbolically (as opposed to describing it in words) is in indexed form. For example consider the set of all open intervals on the line of the form $(-r, r)$ where r is a rational number. It would be impossible to list all of these intervals and equally impossible to list a few to establish a pattern like we did with $\{\frac{1}{2}, \frac{1}{4}, \frac{1}{8}, \cdots\}$. But indexing works perfectly: we can write this set as

$$\{(-r, r) : r \in \mathbf{Q}\}.$$

Note that in this example, the objects in the indexed collection are sets themselves, and we could write

$$\{(-r, r) : r \in \mathbf{Q}\}$$

as

$$\{A_r : r \in \mathbf{Q}\}, \quad \text{where} \quad A_r = (-r, r).$$

It is often the case that the indexed objects in a collection are sets, and this is why we refer to the "collection" of A_λ's instead of the "set" of A_λ's. A "set of sets" sounds strange, but it is often the case that the elements of a set are also sets (remember $\mathcal{P}(X)$?). We will try to refer to a set whose elements are sets as a "collection" or "family" of sets, rather than a set of sets in the hope of minimizing the confusion as much as possible.

3.1. Exercises.

Determine what the following sets are:

1) $\{n \in \mathbf{Z}^+ : m \nmid n$ for all $m \in \mathbf{Z}^+$ with $1 < m < n\}$ (The symbol $m \mid n$ means m divides n (evenly), and $m \nmid n$ means m does not divide n (evenly). Thus $3 \mid 6$ but $3 \nmid 7$.)
2) $\{A_\lambda : \lambda \in \Lambda\}$, where each $A_\lambda = \{\lambda\}$.
3) $\{2r : r \in \mathbf{R}\}$
4) $\{x_A : A \in \mathcal{P}(X)\}$, where $x_A = A$.

You have probably seen the union and intersection of a pair of sets before: if A and B are sets, then the union of A and B, written $A \cup B$, is the set

$$A \cup B = \{x : x \in A \quad \text{or} \quad x \in B\}$$

and the intersection of A and B is the set

$$A \cap B = \{x : x \in A \quad \text{and} \quad x \in B\}.$$

Thus the union of A and B is the set of all elements that belong to at least one of A or B, and the intersection of A and B is the set of all elements that belong to each of A and B. We can generalize these ideas as follows:

3.2. Definition. Let $\mathcal{C} = \{A_\lambda : \lambda \in \Lambda\}$ be a collection of sets indexed by the indexing set Λ.

    1) the **union** of the collection $\mathcal{C}$, denoted $\bigcup \mathcal{C}$ or $\bigcup\{A_\lambda : \lambda \in \Lambda\}$ or $\bigcup_{\lambda \in \Lambda} A_\lambda$, is defined to be

$$\{x : x \in A_\lambda \quad \text{for some} \quad \lambda \in \Lambda\}.$$

    2) The **intersection** of the collection $\mathcal{C}$, denoted $\bigcap \mathcal{C}$ or $\bigcap\{A_\lambda : \lambda \in \Lambda\}$ or $\bigcap_{\lambda \in \Lambda} A_\lambda$, is defined to be

$$\{x : x \in A_\lambda \quad \text{for all} \quad \lambda \in \Lambda\}.$$

This definition really is a generalization of the one given for a pair of sets because the two definitions are exactly the same whenever the collection of sets in Definition 3.2 is a collection of two sets:

3.3. Exercise.

    Show that Definition 3.2 reduces to the one given for $A \cup B$ and $A \cap B$ in case $\mathcal{C}$ is a collection with exactly two elements. Does $\bigcup\{A, B\}$ make sense? [*Hint:* Put $A = A_1$, $B = A_2$. What is Λ?]

Note that, in Definition 3.2, $\bigcup \mathcal{C}$ is defined to be the set of all points that belong to at least one element of $\mathcal{C}$, and we could write

$$\bigcup \mathcal{C} = \{x : \text{there exists an } A \in \mathcal{C} \text{ such that } x \in A\},$$

or

$$\bigcup \mathcal{C} = \{x : \text{there exists a } \lambda \in \Lambda \text{ such that } x \in A_\lambda\}.$$

To keep it straight, remember that $x \in \bigcup \mathcal{C}$ if there is *at least one* member of $\mathcal{C}$ that x belongs to (there may be more than one member of $\mathcal{C}$ that x belongs to when $x \in \bigcup \mathcal{C}$, but the important thing is that there is at least one), and $x \in \bigcap \mathcal{C}$ if x belongs to *every* member of $\mathcal{C}$.

    Note also that no matter how many sets are in the collection $\mathcal{C}$, the elements of both $\bigcup \mathcal{C}$ and $\bigcap \mathcal{C}$ are the same kind of elements that are in the sets that belong to $\mathcal{C}$. For example, if $\mathcal{C} = \{A, B\}$ where $A = \{a, e, 2\}$ and $B = \{a, 1, 2, 3\}$, then $\bigcup \mathcal{C} = \{a, e, 1, 2, 3\}$ and $\bigcap \mathcal{C} = \{a, 2\}$.

3.4. Exercises.
1) Let $\mathcal{C} = \{[-n, n]: n \in \mathbf{Z}^+\}$. What are $\mathsf{U}\mathcal{C}$ and $\mathsf{\cap}\mathcal{C}$?
2) Let $\mathcal{C} = \{(-1/n, 1/n): n \in \mathbf{Z}^+\}$. What are $\mathsf{U}\mathcal{C}$ and $\mathsf{\cap}\mathcal{C}$?
3) Let $\mathcal{C} = \{[-1 + 1/n, 1 - 1/n]: n \in \mathbf{Z}^+\}$. What are $\mathsf{U}\mathcal{C}$ and $\mathsf{\cap}\mathcal{C}$?
4) Let $\mathcal{C} = \{(a, b): a, b \in \mathbf{Q}, a < b\}$. What are $\mathsf{U}\mathcal{C}$ and $\mathsf{\cap}\mathcal{C}$?
5) Let $\mathcal{C} = \{[r, \infty): r \in \mathbf{R}\}$. What are $\mathsf{U}\mathcal{C}$ and $\mathsf{\cap}\mathcal{C}$?
6) Let $\mathcal{C}$ be a collection of subsets of a set X.
   a) State precisely what it means when a point $x \in X$ does *not* belong to $\mathsf{U}\mathcal{C}$.
   b) State precisely what it means when a point $x \in X$ does *not* belong to $\mathsf{\cap}\mathcal{C}$.

The proof of the following theorem is a very straightforward application of the definitions, and the theorem is important.

3.5. Theorem. Let $\mathcal{C} = \{A_\lambda : \lambda \in \Lambda\}$. If $\lambda \in \Lambda$, then

$$\mathsf{\cap}\mathcal{C} \subset A_\lambda \subset \mathsf{U}\mathcal{C}.$$

3.6. Corollary. For any two sets A and B,

$$A \cap B \subset A \subset A \cup B \quad \text{and} \quad A \cap B \subset B \subset A \cup B.$$

Besides union and intersection, another important operation on sets is that of taking complements. Specifically,

3.7. Definition. Let A and B be sets. Then the **complement of A relative to B**, written $B - A$, is defined by

$$B - A = \{x: x \in B \quad \text{and} \quad x \notin A\}.$$

In case we have a **universal set** X (the set of all elements under consideration), the complement of A relative to X is called simply the **complement** of A.

The following exercises are easy (if you remember the discussion of the importance of definitions!). But the results in these exercises are of basic importance.

3.8. Exercises.
1) If $A \subset X$, then $X - (X - A) = A$.
2) If $A, B \subset X$, then $A \subset B$ if and only if $X - A \supset X - B$.
3) If $A, B \subset X$, then $A = B$ if and only if $X - A = X - B$.
4) For $A, B \subset X$, $A - B = A \cap (X - B)$.

Also useful are the *De Morgan Laws** which can be stated roughly as:
The complement of a union is the intersection of the complements, and the
complement of an intersection is the union of the complements. Precisely,

3.9. Theorem. Let $\mathcal{C}$ be a collection of subsets of a set X. Then:
  1) $X - \cup \mathcal{C} = \cap \{X - A : A \in \mathcal{C}\}$
  2) $X - \cap \mathcal{C} = \cup \{X - A : A \in \mathcal{C}\}$.
[*Hint for proof:* Prove (1) by showing directly from the definition that
the two sets are equal; then use (1) with a clever choice for the "$\mathcal{C}$" in (1)
to prove (2).]

3.10. Exercises.
  1) For any two sets A and B,
     a) $A \cap B = A$ if and only if $A \subset B$.
     b) $A \cup B = A$ if and only if $B \subset A$.
     c) $A \cap B = A$ and $A \cup B = A$ if and only if $A = B$.
  2) If $A \subset B$, $C \subset D$ are sets, then
     a) $A \cap C \subset B \cap D$.
     b) $A \cup C \subset B \cup D$.
  3) The distributive properties: Let $\mathcal{C}$ be a collection of subsets of a set
     X, $B \subset X$. Then
     a) $B \cap (\cup \mathcal{C}) = \cup \{B \cap A : A \in \mathcal{C}\}$.
     b) $B \cup (\cap \mathcal{C}) = \cap \{B \cup A : A \in \mathcal{C}\}$.
  4) Let $\mathcal{C}$ be a non-empty collection of sets. Then
     a) $B \cup (\cup \mathcal{C}) = \cup \{B \cup A : A \in \mathcal{C}\}$.
     b) $B \cap (\cap \mathcal{C}) = \cap \{B \cap A : A \in \mathcal{C}\}$.
  5) Show that the hypothesis that $\mathcal{C}$ be non-empty is needed in Problem
     4 but not in Problem 3 above.
  6) Give examples of:
     a) An infinite collection of open intervals that all have a point in
        common whose intersection is not an open interval.
     b) An infinite collection of closed intervals that all have a point in
        common whose union is not a closed interval.

So far, the operations that we have defined on sets (union, intersection,
and complementation) have produced new sets that have the same kind of
elements as the sets that we started with. Specifically, we have seen that A
and B are both subsets of $A \cup B$, $A \cap B$ is a subset of both A and B, and
$A - B$ is a subset of A. Now we want to investigate an important way of

* After Augustus De Morgan (1806–1871).

combining two sets that produces a new set which does not have the same kind of elements as those we start with.

3.11. Definition. Let A and B be sets. The **Cartesian product***** of A and B, written $A \times B$ (read "A cross B"), is defined to be the set

$$A \times B = \{(a, b) : a \in A, b \in B\}$$

and is the set of **ordered pairs** with first coordinate an element of A and second coordinate an element of B.

Two such ordered pairs are equal if and only if they are equal coordinate-wise. In other words, $(a, b) = (c, d)$ if and only if $a = c$ and $b = d$.

For example, if $A = \{0, 1\}$ and $B = \{1, 2, 3\}$, then

$$A \times B = \{(0, 1), (0, 2), (0, 3), (1, 1), (1, 2), (1, 3)\}.$$

Note that the order in the ordered pairs makes a difference and, for example, $(1, 0) \notin A \times B$ with A and B as defined above. Note also that $A \times B$ does not have the same kind of elements as A or B. In particular, there is no containment relationship between A or B and $A \times B$ as there is with union, intersection and complementation.

3.12. Exercises.
1) Let $A = \{2, 4, 6\}$, $B = \{1, 3\}$. What is $A \times B$? What is $B \times A$? Does $A \times B = B \times A$?
2) Let $A = \mathbf{Z}^+$ (the set of positive integers), and let $B = 2\mathbf{Z}^+$ (the set of even positive integers). What is $A \times B$? What is $B \times A$? Are they equal?
3) When does $A \times B = C \times D$? State a theorem and prove it. (Since $A \times B$ and $C \times D$ are sets, this will be a theorem about equality of two sets, so Definition 1.1 will have to be used in the proof, in addition to the definition of Cartesian product.)
4) When does $A \times B = B \times A$? State a theorem and prove it. [*Hint:* See Problem 3 above.]

An important example of the Cartesian product of two sets is the familiar Euclidean plane. You should be absolutely sure that you understand that the plane is just $\mathbf{R} \times \mathbf{R}$, where $\mathbf{R}$ is the set of real numbers.

We can illustrate the Cartesian product of subsets of $\mathbf{R}$ on the plane:

***** After the French mathematician and philosopher Rene Descartes (1596–1650), who invented analytic geometry. In analytic geometry one uses coordinate systems to describe geometric figures algebraically, and the usual coordinate system in the plane is the Cartesian product of two real lines (the "x" and "y" axes), as we will see.

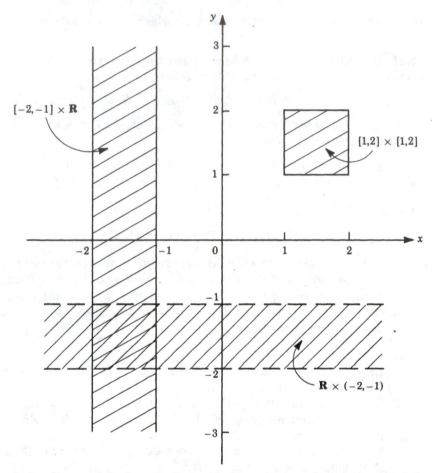

It is often useful when dealing with Cartesian products to pretend that the two sets A and B lie on the line (even if they do not), and then visualize $A \times B$ in the plane. Use this technique (if you need it) in the following exercises, which relate Cartesian product to union, intersection and complementation.

3.13. Exercises.

Decide which of the following statements are true and which are false. Prove those that are true, correct those that are false and prove the corrected version. Let A, B, C, $D \subset X$, where X is some set.

1) $(A \times B) \cup (C \times D) = (A \cup C) \times (B \cup D)$
2) $(A \cup B) \times (C \cup D) = (A \times C) \cup (B \times D)$
3) $(A \times B) \cap (C \times D) = (A \cap C) \times (B \cap D)$
4) $(A \cap B) \times (C \cap D) = (A \times C) \cap (B \times D)$

5) $(A \times B) - (C \times D) = (A - C) \times (B - D)$
6) $(A - B) \times (C - D) = (A \times C) - (B \times D)$.

4. FUNCTIONS

The idea of a function is probably the single most important one in all of mathematics. Fortunately, it is a simple idea: a function from a set X to a set Y is a rule that associates one and only one element of Y with each element of X; it is an unambiguous way to go from X to Y. Formally,

4.1. Definition. Let X and Y be sets. A **function** f from X to Y is a subset of $X \times Y$ such that whenever (x, y_1) and (x, y_2) both belong to the subset f, then $y_1 = y_2$.

This definition may be confusing at first, but a little thought will convince you that it simply makes the intuitive idea of a function mathematically precise: a function f from X to Y associates $y \in Y$ with $x \in X$ if and only if $(x, y) \in f$; and, each $x \in X$ can have only one y associated with it: if y_1 and y_2 are both paired with x by f, then $y_1 = y_2$. Note, however, that we do *not* prohibit x_1 and x_2 from both being paired with the same y, even if $x_1 \neq x_2$.

4.2. Exercises.
1) Which of the following are functions from X to Y, where $X = \{0, 1, 2\}$, $Y = \{a, b, c, d\}$?
   a) $f = \{(0, a), (0, b), (0, c)\}$
   b) $f = \{(0, a), (1, a), (2, c)\}$
   c) $f = \{(0, a), (1, 2)\}$
2) Write the function f from X to Y as a subset of $X \times Y$, when f is defined by the following picture.

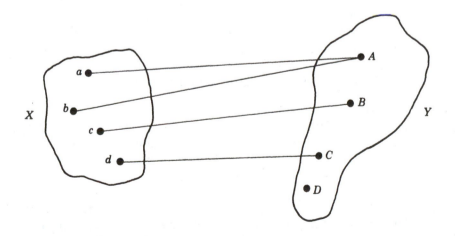

3) Let f be a rule that associates elements of Y with each element of X. State precisely what it means when f is *not* a function from X to Y.

When f is a function from X to Y we write $f:X \rightarrow Y$, or, sometimes, $X \xrightarrow{f} Y$. If $y \in Y$ is the element associated with $x \in X$ by f (i.e., if $(x, y) \in f$), we write $y = f(x)$, read "$y = f$ of x." The set X is called the **domain** of f, and we say that f is **defined on** X, or f is a **function on** X. The set $\{f(x):x \in X\}$ $\subset Y$ is called the **range** of f. A function whose range is a subset of **R** is called a **real-valued** function, and if the domain and range are both subsets of **R**, the function is often called a **real-valued function of a real variable**.

We can often write functions using formulas. For example, if $f:\mathbf{R} \rightarrow \mathbf{R}$ is the function that sends each real number in its domain to the square of that real number in its range, we can write "$f:\mathbf{R} \rightarrow \mathbf{R}$ by $f(x) = x^2$."

If $A \subset X$, then the **image of** A **under** f, written $f(A)$ and read "f of A," is the subset of Y defined by

$$f(A) = \{y \in Y : y = f(x) \quad \text{for some} \quad x \in A\},$$

which is the same as

$$\{f(x) : x \in A\}.$$

Thus the range of f is the image of its domain. Also, for $B \subset Y$, the **preimage** or **inverse image** of B, written $f^{-1}(B)$ and read "f inverse of B," is the subset of X defined by

$$f^{-1}(B) = \{x \in X : f(x) \in B\}.$$

You should convince yourself that $f^{-1}(B)$ is defined for all $B \subset Y$ and that f^{-1} is a function from $\mathcal{P}(Y)$ to $\mathcal{P}(X)$. Sometimes we can use f^{-1} to induce a function from Y to X, and for that, we need the following definition.

4.3. Definition. Let $f:X \rightarrow Y$.
  1) f is **1-1** (or **injective**) if whenever $f(x_1) = f(x_2)$ then $x_1 = x_2$.
  2) f is **onto** (or **surjective**) if $f(X) = Y$, i.e., if for all $y \in Y$ there is an $x \in X$ such that $f(x) = y$.
  3) f is **bijective** if it is both injective and surjective.

4.4. Exercises.
  1) State precisely what it means when a function $f:X \rightarrow Y$ is not 1-1.
  2) State precisely what it means when a function $f:X \rightarrow Y$ is not onto.
  3) Let $f:\mathbf{R} \rightarrow \mathbf{R}$ be defined by $f(x) = x^2$. Graph the function f in the plane and show its domain and range. Compute $f^{-1}(\{0\})$ and $f^{-1}(\{4\})$. Is f 1-1?
  4) Let $f:\mathbf{R}^+ \rightarrow \mathbf{R}$ by $f(x) = x^2$. ($\mathbf{R}^+$ is the set of positive real numbers.) Graph f in the plane and show its domain and range. Compute

$f^{-1}(\{0\})$, $f^{-1}(\{4\})$ and $f^{-1}(\{r\})$ when $r > 0$. What is $f^{-1}(\{r\})$ when $r \leq 0$? Is f 1-1?

5) Conjecture a theorem that says whether or not a function f from X to Y is 1-1 in terms of the number of points in $f^{-1}(\{y\})$ for each $y \in Y$. Prove your theorem.

6) Compute $f^{-1}([1, 2])$ for the function f defined in exercise (3) above. Draw it on your graph.

Compute $f^{-1}([1, 2])$ for the function f defined in Problem 4 above. Draw it on your graph.

When $f: X \to Y$ is both 1-1 and onto, then for each $y \in Y$, $f^{-1}(\{y\}) = \{x\}$ if and only if $f(x) = y$. The convention in this case is to drop the braces and write

$$f^{-1}(y) = x \quad \text{if and only if} \quad f(x) = y.$$

Thus when $f: X \to Y$ is bijective, we can think of f^{-1} as the function from Y to X that "undoes what f does," in the sense that $f^{-1}(y) = x$ if and only if $f(x) = y$. The function $f^{-1}: Y \to X$ defined in this way is called the **inverse** of the function f.

Notice that we are using the symbol f^{-1} in two different ways. When f is any function from X to Y and B is any subset of Y, then our original definition of f^{-1} says that

$$f^{-1}(B) = \{x \in X : f(x) \in B\},$$

and in this case, f^{-1} is a function from $\mathcal{P}(Y)$ to $\mathcal{P}(X)$, and not a function from Y to X. But in the particular case that $f: X \to Y$ is 1-1 and onto, we can think of f^{-1} as the function from Y to X defined by $f^{-1}(y) = x$ if and only if $f(x) = y$, and we call this function f^{-1} the inverse of the function f. To further complicate things, we often call f^{-1} the inverse of f even when f is not 1-1 and onto, but when we do this, we have to keep in mind that f^{-1} may not be a function from Y to X.

The following theorem tells how functions and their inverses behave with respect to unions and intersections.

4.5. Theorem. Let $f: X \to Y$, let $\mathcal{C}$ be a collection of subsets of X and let $\mathcal{D}$ be a collection of subsets of Y. Then

1) $f(\cup \mathcal{C}) = \cup\{f(A) : A \in \mathcal{C}\}$
2) $f(\cap \mathcal{C}) \subset \cap\{f(A) : A \in \mathcal{C}\}$
3) $f^{-1}(\cup \mathcal{D}) = \cup\{f^{-1}(B) : B \in \mathcal{D}\}$
4) $f^{-1}(\cap \mathcal{D}) = \cap\{f^{-1}(B) : B \in \mathcal{D}\}.$

There is one more definition that we will need later.

4.6. Definition. Let $f: X \to Y$ and $g: X \to Y$. Then $f = g$ if and only if $f(x) = g(x)$ for every $x \in X$.

Thus two functions with the same domain and range are the same if and only if they do exactly the same thing to every point of the domain.

4.7. Exercises.

1) Let $f: X \to Y$, $A \subset X$ and $B \subset Y$. Then
   a) $A \subset f^{-1}(f(A))$
   b) $f(f^{-1}(B)) = B \cap f(X)$
   c) Give an example to show that the containment in (a) above can be proper. (A subset S of a set X is a *proper* subset if $S \subset X$ but $S \neq X$.)
   d) Give a condition on f that will make $A = f^{-1}(f(A))$ for every $A \subset X$, and prove that your condition is right.
   e) Give a condition on f that will make both $A = f^{-1}(f(A))$ for every $A \subset X$ and $f(f^{-1}(B)) = B$ for every $B \subset Y$, and prove that your condition is right.

2) Is the following statement correct? If $x \in X$ and $f: X \to Y$ then $x = f^{-1}(f(x))$.

3) Give an example to show that the containment in Theorem 4.5 (2) can be proper. (A $\mathcal{C}$ consisting of only two sets will do, and it might be helpful to look at the real line.)

5. COMPOSITION OF FUNCTIONS

Consider the following situation:

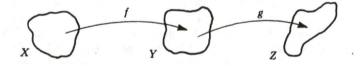

Here f is a function from X to Y and g is a function from Y to Z. The question is: can we use f and g in some way to get a function from X to Z? Obviously the answer is yes because we can go from X to Z in "steps": first we can go from X to Y by using f and then we can transfer to g and go on to Z by using the function g. The resulting function that takes X into Z is called the *composition* of f and g and is denoted by $g \circ f$. Formally,

5.1. Definition. Let $f: X \to Y$, $g: Y \to Z$. Then the **composition** of g and f is the function $g \circ f: X \to Z$ defined by

$$g \circ f(x) = g(f(x)).$$

5.2. Exercises.

1) Let $f: \mathbf{R} \to \mathbf{R}$ and $g: \mathbf{R} \to \mathbf{R}$ be defined by $f(x) = x^2$, $g(x) = x^3$. What is $f \circ g$? What is $g \circ f$?

2) Let $f:\mathbf{R} \to \mathbf{R}$ and $g:\mathbf{R} \to \mathbf{R}$ be defined by $f(x) = 2x + 1, g(x) = x^2$. What is $f{\circ}g$? What is $g{\circ}f$?

3) Let $f:\mathbf{R} \to \mathbf{R}$ and $g:\mathbf{R} \to \mathbf{R}$ be defined by $f(x) = x + 1, g(x) = x - 1$. What is $f{\circ}g$? What is $g{\circ}f$?

A useful way of looking at composition of functions is by using diagrams. Suppose that $f:X \to Y$ and $g:Y \to Z$. A diagram of this situation might look like:

$$X \xrightarrow{\ f\ } Y$$
$$Z \nearrow g$$

It is clear from the diagram that we can go from X to Z by going through Y, and that the resulting function is the function f followed by the function g, or simply the function $g{\circ}f$. Putting this on the diagram we get

$$X \xrightarrow{\ f\ } Y$$
$$g{\circ}f \searrow \quad \nearrow g$$
$$Z$$

The completed diagram is said to be *commutative* because no matter how you go from X to Z, the result is the same: for each point $x \in X$,

$$g(f(x)) = g{\circ}f(x).$$

Commutative diagrams can be generalized to more than three spaces, and they play an important role in mathematics. For example, consider the diagram

$$W \xrightarrow{\ f\ } X$$
$$\quad\quad \downarrow g$$
$$Z \xleftarrow{\ h\ } Y$$

The obvious way to complete this diagram so that the resulting completed diagram is commutative is

$$W \xrightarrow{\ f\ } X$$
$$F \downarrow \quad\quad \downarrow g$$
$$Z \xleftarrow{\ h\ } Y$$

where $F = h{\circ}g{\circ}f$.

Suppose, though, that we are told that the diagram

is commutative. We can immediately conclude that $g \circ f = F \circ G$.

We can use commutative diagrams to visualize an important function.

5.3. Definition. Let X be a set. The function $\mathrm{id}_X : X \to X$ defined by $\mathrm{id}_X(x) = x$ is called the **identity** function on X.

Consider the following diagram.

$$
\begin{array}{ccc}
X & \xrightarrow{\;f\;} & X \\
\mathrm{id}_X \downarrow & & \downarrow \mathrm{id}_X \\
X & \xrightarrow{\;f\;} & X
\end{array}
$$

This diagram is commutative if and only if $\mathrm{id}_X \circ f = f \circ \mathrm{id}_X$, a fact which you can readily verify. In fact, it is trivial to show that both $\mathrm{id}_X \circ f = f$ and $f \circ \mathrm{id}_X = f$, so that the following extended diagram is also commutative:

$$
\begin{array}{ccc}
X & \xrightarrow{\;f\;} & X \\
\mathrm{id}_X \downarrow & \searrow^{f} & \downarrow \mathrm{id}_X \\
X & \xrightarrow{\;f\;} & X
\end{array}
$$

Thus $\mathrm{id}_X \circ f = f \circ \mathrm{id}_X = f$ and the function $\mathrm{id}_X : X \to X$ plays the same role relative to composition of functions from X to X as 0 does relative to addition of real numbers and as 1 does relative to multiplication of (non-zero) real numbers: they are the *identity elements* of their respective operations.

5.4. Exercises.
1) If $f : X \to X$ is 1-1 and onto (so f^{-1} can be thought of as a function from X to X) then $f \circ f^{-1} = f^{-1} \circ f = \mathrm{id}_X$. Thus f^{-1} really is the inverse of f relative to composition of functions because the composition of f with f^{-1} in either order gives the identity function. This is exactly the same idea as $n(1/n) = (1/n)n = 1$ or $n + (-n) = -n + n = 0$ for positive integers n.
2) Let $f : X \to Y$ and $g : Y \to Z$. Prove that if f and g are both 1-1, then $g \circ f$ is 1-1. Is the converse true?
3) Let $f : X \to Y$ and $g : Y \to Z$. Prove that if f and g are both onto then $g \circ f$ is onto. Is the converse true?

chapter two

Infinite Sets and Transfinite Numbers

The concept of infinity is a fascinating and important one. In this chapter, we will investigate some infinite sets and the transfinite numbers which are used with them.

1. BASIC IDEAS

1.1. Definitions.
   a) A **1-1 correspondence** between two sets is a bijective (1-1 and onto) function between them.
   b) Two sets have the **same cardinality** if there is a 1-1 correspondence between them.

According to this definition, the two sets $\{a, e, i, o, u\}$ and $\{0, 1, 2, 3, 4\}$ have the same cardinality: an example of a bijective function between them that proves it might be one that sends a to 0, e to 1, i to 2, o to 3, and u to 4. Clearly, "same cardinality" has something to do with "same number of elements," and it will not hurt at all if you think of it in this way (in fact, it will probably help keep things clear, but keep in mind that "same cardinality as" has a precise definition when you need it in a proof).

It is when we deal with infinite sets that the idea of two sets having the same cardinality becomes really fascinating, because with infinite sets we can have things like a *proper* subset of a set that has the same number of elements as the set itself (try this with finite sets—it won't happen). It *always* happens with infinite sets, and we could follow Dedekind* and define what it means for a set to be infinite by saying that a set is infinite if and

---

* Richard Dedekind (1831–1916), one of the pioneers in the mathematical theory of the infinite.

only if it has a proper subset with the same cardinality as itself. We choose to be a little less startling though, and give a definition that determines whether a set is infinite or not by comparing it with a kind of "base" set, the set of positive integers.

1.2. Definition. A set S is **finite** if it is empty or if there is a positive integer N such that S can be put into a 1-1 correspondence with the set $\{1, 2, \cdots, N\}$; S is **infinite** if it is not finite.

Clearly, our definition says that the set of positive integers itself is infinite. It is also infinite in the Dedekind sense. To show it, we have to show that it has a proper subset with the same cardinality as itself. The even positive integers (denoted $2\mathbf{Z}^+$) will work: define $f:\mathbf{Z}^+ \to 2\mathbf{Z}^+$ by $f(n) = 2n$, and show that f is a 1-1 correspondence between the two sets.

The fact that there are exactly as many even positive integers as there are all positive integers is probably surprising. Intuition might say that since there are the same number of even positive integers as there are odd positive integers, then when the evens and odds are combined to form the set of all positive integers, there should be twice as many positive integers as even positive integers. This intuition would be wrong, as we have just seen, so you can see that intuition will have to be used with care when dealing with infinite sets.

We have defined what it means for two sets to have the same cardinality, but we have not said what the cardinality of a set is. In the case of finite sets, the cardinality of a set is nothing but the number of elements in the set; more precisely, the cardinality of S is n if and only if S can be put into a 1-1 correspondence with the set $\{1, 2, 3, \cdots, n\}$. This idea could easily be extended to infinite sets if we had some infinite numbers to match our sets up with; they exist and we will see them later, but we do not have them now. For the moment, we will assume that "the number of elements in a set" is a phrase that always makes sense (it does) and to make the discussion easier, we will call the number of elements in a set S the **cardinality** of S and denote it by the symbol $|S|$.

Thus we have $|A| = |B|$ if and only if there is a 1-1 correspondence between A and B. We say that a set S is **countable** if it is finite or if $|S| = |\mathbf{Z}^+|$. Sometimes, for emphasis, if a set is countable and infinite, we will say that it is **countably infinite.**

1.3. Exercises.
1) Show that the set of odd positive integers is countably infinite.
2) Show that the set $\mathbf{Z} = \{\cdots, -3, -2, -1, 0, 1, 2, 3, \cdots\}$ of all integers is countably infinite.
3) Show that for any positive integer n, the number of subsets of an n element set is 2^n. [*Hint:* Use induction.]

The symbol used for the cardinality of a countably infinite set is $\aleph_0$ (aleph sub zero—aleph is the first letter of the Hebrew alphabet). The symbol $\aleph_0$ denotes a number, just like 3 or 8; it is our first example of a number giving the size of an infinite set, and it is a **transfinite** cardinal number. (In general, a number (finite or infinite) is a **cardinal number** if it gives the size (the number of elements) of a set.) Writing $\aleph_0 = |\mathbf{Z}^+|$ is the same idea as writing $3 = |\{0, 1, 2\}|$.

We have so far that

$$2\mathbf{Z}^+ \subsetneqq \mathbf{Z}^+ \subsetneqq \mathbf{Z}$$

and yet

$$|2\mathbf{Z}^+| = |\mathbf{Z}^+| = |\mathbf{Z}| = \aleph_0,$$

which, when you think about it, is quite a departure from the case of finite sets. Here are three sets, each properly contained in the next, and yet all three of them have exactly the same number of elements—they are all countably infinite. There are even "larger" sets than $\mathbf{Z}$ which are also countably infinite (larger in the sense of containment, of course—not larger in the sense of having a greater number of elements). Before we can deal with them, though, we need the following definition.

1.4. Definition. For two sets A and B, $|A| \leq |B|$ if there is a 1-1 function from A into B, and $|A| < |B|$ if there is a 1-1 function from A into B but there is *not* a 1-1 function from A *onto* B.

The use of the symbol "$\leq$" in this definition is very suggestive, and it probably seems obvious that if $|A| \leq |B|$ and $|B| \leq |A|$ then $|A| = |B|$. This is true, but it is surprisingly difficult to prove. It is called the Schröder-Bernstein theorem,* and you can find a proof of it in Wilder [18]. We will accept it and use it without proving it, and for later reference, we state it formally as a theorem. (Don't try to prove it.)

1.5. Theorem (Schröder-Bernstein). For two sets A and B, $|A| = |B|$ if and only if both $|A| \leq |B|$ and $|B| \leq |A|$. In other words, $|A| = |B|$ if and only if there exists a 1-1 function from A into B and there also exists a 1-1 function from B into A.

As an immediate corollary of the Schröder-Bernstein theorem we have the following "squeeze" theorem, which is often very useful. (You should prove this corollary.)

1.6. Corollary. If $A \subset B \subseteq C$ are sets and $|A| = |C|$, then $|A| = |B| = |C|$.

---

* After Ernst Schröder (1841–1902) and Felix Bernstein (1878–1956) who proved it independently.

Use the squeeze theorem and the given outline to prove the following theorem (which we will then use to prove that the set of rational numbers is countable).

1.7. Theorem. The union of a countable collection of countable sets is countable, i.e., if $\mathcal{C} = \{A_\lambda : \lambda \in \Lambda\}$ with each $|A_\lambda| \leq \aleph_0$ and $|\Lambda| \leq \aleph_0$, then $|\mathcal{UC}| \leq \aleph_0$.

Outline of Proof: Build a checkerboard with countably infinitely many squares in two directions. Then count the squares in a clever way, thus inducing a 1-1 correspondence between the positive integers and the squares. Now put your sets on the board (one set to a row) and with the aid of the squeeze theorem (Corollary 1.6), conclude that the union of your sets is countable. (The way that the squeeze theorem is to be used here is a little subtle. Counting the squares in the checkerboard shows that the number of squares is countable. There are then two problems that can occur: when you put your sets on the board as directed, you may not use up all of the squares —some of your sets might be finite; secondly, some of your sets may have elements in common which would be counted separately when you put the sets on the board, but would only be counted once in the union of your sets. Even so, the number of elements in the union of your sets is no larger than the number of squares (Why?), so if any one of your sets is infinite or if Λ is infinite (or both), the squeeze theorem may have to be used (Why?). (Can a countable union of countable sets ever be finite?).)

Since the set of rational numbers can be written as the union of a countable collection of countable sets, the following corollary is immediate from Theorem 1.7. To prove it, you should show how the rationals can be written as the countable union of countable sets.

1.8. Corollary. The set $\mathbf{Q}$ of rational numbers is countable.

The set of rationals should seem *gigantic* relative to the set of positive integers, because *all* fractions (with numerator and denominator both integers) are rational numbers, but only those fractions which are positive and have denominator 1 (in lowest terms) are positive integers. Yet we have shown that the number of rational numbers is the same as the number of positive integers! This is a very striking example of a set ($\mathbf{Q}$) with a proper subset ($\mathbf{Z}^+$) with the same cardinality as itself. The set of integers is in a sense "twice as big" as the set of positive integers; the set of rational numbers is "infinitely as big", but all three of them have the same cardinality.

There is another set which looks even more gigantic than the set of rational numbers, but we will show that it too is countable. This is the set of algebraic numbers.

1.9. Definition. A (real) number is **algebraic** if it is a root of a poly-nomial with integer coefficients.

For example, the numbers 2, 2/3 and $\sqrt{2}$ are algebraic because they are roots of the polynomial equations $x - 2 = 0$, $3x - 2 = 0$ and $x^2 - 2 = 0$, respectively.

Clearly every rational number is algebraic (prove it), and there are "lots" of algebraic numbers that are not rational (What are some?). But the set of algebraic numbers is countable. To prove it, use Theorem 1.7 to observe that for each positive integer n, the set of polynomials of degree n with integer coefficients is countable. (The *degree* of a polynomial is the highest power of the unknown that appears with a non-zero coefficient.) Then use Theorem 1.7 again to show that the set of *all* polynomials with integer coefficients is also countable. Finally, observe that the number of roots of a given polynomial is finite (and therefore countable), and use Theorem 1.7 once again to get the result.

1.10. Theorem. The set of algebraic numbers is countable.

Before jumping to any false conclusions about the countability of all sets of real numbers, let us prove the following theorem (which says that, com-pared to the infinite set $\mathbf{Z}^+$, $\mathbf{R}$ is *super* infinite). It also shows that there are different degrees of infinity—two infinite sets need not be the same size—one can be larger than the other, even though both are infinite.

1.11. Theorem. The set $\mathbf{R}$ is not countable.

Outline of Proof: Suppose that $\mathbf{R}$ is countable. Then there is a 1-1 cor-respondence f between $\mathbf{Z}^+$ and $\mathbf{R}$, say $f(1) = r_1 = N_1.d_{11}d_{12}d_{13} \cdots$, $f(2) = N_2.d_{21}d_{22}d_{23} \cdots$, where we have written the real numbers r_1, r_2, etc. as deci-mals. Now show that the function f cannot be onto (as it is assumed to be) by constructing a real number that is different from every $f(n)$. Do this by making the new number differ from each $f(n)$ in the n-th decimal place.

A real number that is not algebraic is called **transcendental**. The following corollary is immediate from Theorems 1.11 and 1.7.

1.12. Corollary. The set of transcendental numbers is not countable.

This corollary deserves some comment because the only two transcen-dental numbers that most people ever see (if they see any at all) are π, the ratio of the circumference of a circle to its diameter, and e, the base for natural logarithms. (We should also remark that it is very hard to prove that π and e are transcendental.) The corollary says, though, that there are fantastically many different transcendental numbers—in fact, there are as many transcendental numbers as there are all real numbers combined.

(Note that the corollary is an example of an *existence* statement: it says that lots of transcendental numbers exist, but it does not say what they are.)

The symbol used for the cardinality of the set of real numbers is c. Since the function $f(n) = n$ sends $\mathbf{Z}^+$ 1-1 *into* $\mathbf{R}$ and the proof of Theorem 1.11 shows that there cannot be a 1-1 function from $\mathbf{Z}^+$ *onto* $\mathbf{R}$, we have $|\mathbf{Z}^+| < |\mathbf{R}|$. This result is important and we state it as a theorem.

1.13. Theorem. $\aleph_0 < c$.

Recall that the number of subsets of an n element set is 2^n. We copy this and define the symbol $2^{\aleph_0}$ to be the number of subsets of the positive integers. In other words, we define the symbol $2^{\aleph_0}$ by $2^{\aleph_0} = |\mathcal{P}(\mathbf{Z}^+)|$.

The following theorem relates the two numbers c and $2^{\aleph_0}$, and its proof is probably the most difficult that we have encountered so far. The basic idea of the proof is to show that $c = 2^{\aleph_0}$ (i.e., that the number of real numbers is the same as the number of subsets of the positive integers) by making a decimal out of each subset of $\mathbf{Z}^+$. It is easier if you first observe that $|\mathbf{R}| = |[0, 1)| = |(0, 1) \cup \{0\}|$, which can be done geometrically by projecting on the following picture:

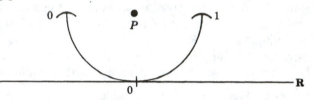

where we have bent the interval $(0, 1)$ into a semicircle, and placed it tangent to $\mathbf{R}$ with $\frac{1}{2}$ on $(0, 1)$ touching 0 on $\mathbf{R}$. Then show that $|[0, 1)| = |\mathcal{P}(\mathbf{Z}^+)|$ by making decimals out of subsets of $\mathbf{Z}^+$, and subsets of $\mathbf{Z}^+$ out of decimals. With some care and ingenuity, this will give you 1-1 functions from $\mathbf{Z}^+$ into $\mathbf{R}$ and from $\mathbf{R}$ into $\mathbf{Z}^+$, and you can then use the Schröder-Bernstein theorem.

1.14. Theorem. $c = 2^{\aleph_0}$.

Since $\aleph_0 < c = 2^{\aleph_0}$, a natural question presents itself: are there any numbers between $\aleph_0$ and $2^{\aleph_0}$? In terms of sets, this question becomes: are there any infinite subsets of the real line which are not countable, but which have cardinality less than that of the line itself? No one has ever exhibited such a set, and the assumption that there are none is known as the **continuum hypothesis.** Another way of stating the continuum hypothesis is: there are no numbers between $\aleph_0$ and $2^{\aleph_0}$.*

* For a good discussion of the continuum hypothesis, see Wilder [18]. We should remark that the continuum hypothesis cannot be proved or disproved using the usual axioms of set theory, and that assuming its *negation* is just as valid mathematically as assuming the continuum hypothesis itself. We will have no need for the continuum hypothesis in this course.

1.15. Exercise.

As we saw above, the set of *all* subsets of $\mathbf{Z}^+$ is uncountable. However, the set of *finite* subsets of $\mathbf{Z}^+$ is countable. To show it, for each finite subset $S = \{n_1, n_2, \cdots, n_k\}$ of $\mathbf{Z}^+$ with $n_1 < n_2 < \cdots < n_k$, associate the integer $2^{n_1} 3^{n_2} 5^{n_3} \cdots p_k^{n_k}$, where p_k is the k-th prime.

2. CARDINAL NUMBERS LARGER THAN c.

Recall that $c = |\,\mathcal{P}(\mathbf{Z}^+)\,|$ and that we proved that $c = 2^{\aleph_0}$, where $\aleph_0 = |\,\mathbf{Z}^+\,|$. In other words, we have proved that $|\,\mathcal{P}(\mathbf{Z}^+)\,| > |\,\mathbf{Z}^+\,|$. We can generalize this in the following theorem and show that for any set X, $|\,\mathcal{P}(X)\,| > |\,X\,|$. In particular, then, when $X = \mathcal{P}(\mathbf{Z}^+)$, this gives $|\,\mathcal{P}(\mathcal{P}(\mathbf{Z}^+))\,| > |\,\mathcal{P}(\mathbf{Z}^+)\,|$, and when $X = \mathcal{P}(\mathcal{P}(\mathbf{Z}^+))$, $|\,\mathcal{P}(\mathcal{P}(\mathcal{P}(\mathbf{Z}^+)))\,| > |\,\mathcal{P}(\mathcal{P}(\mathbf{Z}^+))\,|$, etc. In this way, we obtain larger and larger cardinal numbers (so c is not so super infinite after all!). Letting $2^{|X|} = |\,\mathcal{P}(X)\,|$, we have

$$\aleph_0 < 2^{\aleph_0} = c < 2^c < 2^{2^c} < \cdots, \quad \text{forever.}$$

The outline given for the proof of the theorem that $|\,X\,| < |\,\mathcal{P}(X)\,|$ is taken from a proof of Cantor* as it appears in Kelley ([9], p. 276).

2.1. Theorem. For any set X, $|\,X\,| < |\,\mathcal{P}(X)\,|$.

Outline of Proof: Clearly $|\,X\,| \leq |\,\mathcal{P}(X)\,|$ (Why?). To show that $|\,X\,| < |\,\mathcal{P}(X)\,|$, suppose that they are equal. Then there is a function f that takes X 1-1 *onto* $\mathcal{P}(X)$. Show that the set $\{x \in X : x \notin f(x)\}$ is in $\mathcal{P}(X)$ but is not in the image of X under f, thus contradicting the fact that f is onto.

Recall that the continuum hypothesis says that there are no numbers between $\aleph_0$ and $2^{\aleph_0}$. The **generalized continuum hypothesis** (GCH) extends this to larger cardinal numbers by saying that if $\mathfrak{m}$ is any transfinite cardinal number, then there are no numbers between $\mathfrak{m}$ and $2^{\mathfrak{m}}$ (where, as above, $2^{\mathfrak{m}} = |\,\mathcal{P}(X)\,|$ when $|\,X\,| = \mathfrak{m}$.) Thus the GCH says that there are no numbers between c and 2^c, none between 2^c and 2^{2^c} etc.†

3. ORDINAL NUMBERS

Up to this point we have been concerned only with the size of sets and have not cared at all what the sets really look like. For example, we have shown that the set of positive integers and the set of positive rational numbers are both the same size—they are both countably infinite. But these two sets actually look *very* different: the integers are nicely spaced along the

---

* Georg Cantor (1845–1918) was the founder of the mathematical theory of infinite sets and transfinite numbers. He is responsible for almost all of the material that we have covered (and will cover) on infinite sets.

† Like the continuum hypothesis, the generalized continuum hypothesis cannot be proved or disproved using the ordinary axioms of set theory, and neither can its negation.

line (there is always a distance of at least one unit between any pair of integers), while we can always find rational numbers as close together as we want.

In this section we will be concerned with what sets look like as well as how big they are. To do this completely rigorously would delay our getting into topology much too long, so we choose to describe the situation intuitively in order to get a working knowledge of the ideas involved.*

To begin with, consider the positive integer 5. We usually think of 5 as a cardinal number and use it to describe the size of certain sets—the number of sides of a pentagon, for example, or the number of fingers on one hand. But, if we wanted to, we could think of the integer 5 as a *set* (of all things); indeed, if we assume that the integer 0 has been defined in terms of sets (we might think of 0 as being the empty set $\emptyset$), then we could put

$$1 = \{0\}, \qquad 2 = \{0, 1\}, \qquad 3 = \{0, 1, 2\}, \qquad 4 = \{0, 1, 2, 3\}$$

and, finally,

$$5 = \{0, 1, 2, 3, 4\}.$$

Thus 5 can be thought of as the set of non-negative integers that precede it (where we insist on order: $0 < 1 < 2 < 3 < 4$). Clearly, the integer 6 could now be written as

$$6 = \{0, 1, 2, 3, 4, 5\}.$$

It is important to notice that when we go from $5 = \{0, 1, 2, 3, 4\}$ to $6 = \{0, 1, 2, 3, 4, 5\}$, we adjoin only *one* new element to the set 5 to get the set 6, namely, the number 5; we do *not* adjoin five new elements to the set 5 to get the set 6. (It is true that we adjoin a set consisting of five elements to the set 5 to get the set 6, but this is only one set. This is consistent with adding one unit to the *integer* 5 to get the *integer* 6; we adjoin one element to the *set* 5 to get the *set* 6.) Notice also that the number of elements in the set $5 = \{0, 1, 2, 3, 4\}$ is right: it is true (as it ought to be) that $|5| = 5$.

When we think of 5 as the ordered set of non-negative integers that precede it (i.e., as $5 = \{0, 1, 2, 3, 4\}$), we are thinking of 5 as an **ordinal number**; when we think of it as the number of elements in a set without regard to what the set is or how it looks (except for its size), we are thinking of 5 as a **cardinal number**. The two ideas are related of course: $|5| = 5$.

Obviously there is nothing special about the integer 5 and our idea of regarding 5 as a set can easily be generalized. Indeed if n is any positive integer, we can think of n as an ordinal number and write

$$n = \{0, 1, 2, 3, \cdots, n - 1\},$$

* A rigorous (and very interesting) treatment can be found in Wilder's book [18].

where $n - 1 = \{0, 1, 2, 3, \cdots, n - 2\}$, $n - 2 = \{0, 1, 2, 3, \cdots, n - 3\}$, and so on, and $0 = \emptyset$. (If this doesn't remind you of mathematical induction, it ought to.) We can also think of the positive integer n as a cardinal number, and we have $|n| = n$.

Non-negative integers are *finite* ordinal numbers. Before we can discuss ordinals any further—in particular, before we can examine the fascinating idea of infinite ordinal numbers—there is one more thing that needs to be mentioned. We have been sticking pretty close to familiar notation and will get in trouble if we don't generalize a little. For example, we have said that we can think of the ordinal number 5 as the ordered set of non-negative integers that precede it, and have written

$$5 = \{0, 1, 2, 3, 4\}.$$

But we did not *define* the ordinal number 5 to be this set; indeed, we did not define it at all, and will not, because a rigorous definition turns out to be too complicated for our purposes. The problem is that $\{-3, -2, -1, 0, 1\}$ and $\{\frac{1}{4}, \frac{1}{3}, \frac{1}{2}, 0, 3\}$ are both the ordinal number 5 when they are given their usual ordering, and you can think of many more such examples. An ordinal number depends only on *order* and *size;* what the elements are in the set which is the ordinal number is totally irrelevant. Thus if a, b, c, d, e are any objects (not necessarily numbers) which are ordered by $a < b < c < d < e$, then the set $\{a, b, c, d, e\}$ *is* the ordinal number 5, but the sets $\{b, a, c, d, e\}$ and $\{e, d, c, b, a\}$ for example, are *not*.

Thus we could say that $\{0, 1, 2, 3, 4\}$ is a convenient representative of the collection of sets that are the ordinal number 5, and any set which can be put into a 1-1 correspondence with $\{0, 1, 2, 3, 4\}$ such that order is preserved by the correspondence is also the ordinal number 5. Certainly this idea can be generalized to any finite ordinal number n, and writing

$$n = \{0, 1, 2, 3, \cdots, n - 1\}$$

is convenient, but we have to remember that any set that can be put in a 1-1 correspondence with $\{0, 1, 2, 3, \cdots, n - 1\}$ such that order is preserved by the correspondence is also the ordinal number n. We will call $n = \{0, 1, 2, 3, \cdots, n - 1\}$ the *canonical form* of the ordinal number n.

Writing ordinal numbers in canonical form makes very clear the order relation between non-negative integers. For example, everyone knows that $4 < 5$, but not everyone can give a good reason why. We can now say that $4 < 5$ because, in canonical form, $4 = \{0, 1, 2, 3\}$ is a proper subset of $5 = \{0, 1, 2, 3, 4\}$. In general, for non-negative integers (finite ordinal numbers) n and m, $n < m$ if and only if $n \in m$ when both are written in canonical form. It should also be clear that two non-negative integers m and n are equal as finite ordinal numbers if and only if they are equal as finite cardinal numbers. In other words,

$$\{0, 1, 2, 3, \cdots, m - 1\} = \{0, 1, 2, 3, \cdots, n - 1\}$$

if and only if $n = m$. This may seem too obvious to mention, but just wait!
Consider now the infinite ordered set

$$\mathbf{Z}^+ \cup \{0\} = \{0, 1, 2, 3, \cdots\}$$

consisting of *all* non-negative integers. If there were a largest positive integer N, then this set would represent an ordinal number which we might denote by $N + 1$ (Why?). There is no largest integer N, though (Why not?), but the set $\mathbf{Z}^+ \cup \{0\}$ looks like it ought to represent an ordinal number anyway—after all, it consists of integers ordered in the right way: $0 < 1 < 2 < \cdots$. We can look at the problem from the other direction. We cannot hope to have a largest integer N, but it might be possible to have a smallest ordinal number that is *not* an integer. Indeed it *is* possible, and one way to think of this smallest non-integer ordinal number is as the set $\{0, 1, 2, 3, \cdots\}$ of all non-negative integers ordered in the usual way. This makes sense: the set contains every non-negative integer so can be thought of as being larger than every such integer, and it contains nothing but non-negative integers, so it is certainly the most economical (the smallest) set that contains them all. When we think of it as an ordinal number we call this set ω (small Greek omega; sometimes ω_0 is used instead).

The ordinal number ω is then the smallest infinite (transfinite) ordinal number. It is the smallest ordinal number which is larger than every finite ordinal number (recall that a finite ordinal number is just a non-negative integer), and, in canonical form,

$$\omega = \{0, 1, 2, 3, \cdots, n, \cdots : n \in \mathbf{Z}^+\}.$$

Thus the smallest infinite ordinal number is (in canonical form) just the set of all finite ordinal numbers, ordered in the usual way.

Thus we have ordinal numbers

$$0, 1, 2, 3, \cdots, \omega$$

where each is larger than any one that comes before it (because each is equal, in canonical form, to the set of all of those that come before it).

As with the finite ordinal numbers, thinking of ω in canonical form (i.e., as being equal to the set of all ordinal numbers that precede it) is convenient, but we have to be aware that any set which can be put into a 1-1 correspondence with $\{0, 1, 2, 3, \cdots\}$, such that order is preserved by the correspondence, is also the ordinal number ω.

3.1. Exercises.

Which of the following sets is the ordinal number ω when given its usual ordering? If any of these sets is not ω, why isn't it?

1) The set of positive integers that are prime: $\{2, 3, 5, 7, 11, \cdots\}$.
2) The set of positive even integers: $\{2, 4, 6, 8, \cdots\}$.
3) The set of negative integers $\{-1, -2, -3, \cdots\}$.
4) The set $\{2^n : n \in \mathbf{Z}^+\}$.
5) The set $\{0, 2, 3, 4, 5, \cdots\}$.
6) The set $\{1/n : n \in \mathbf{Z}^+\}$.

Calling $\{0, 1, 2, 3, \cdots, n, \cdots : n \in \mathbf{Z}^+\}$ with its usual order the canonical form of the ordinal number ω enables us to say that an ordinal number n is less than ω if $n \subsetneqq \omega$ when both are represented in canonical form. Thus any finite ordinal number (non-negative integer) is less than ω.

Sometimes when we really want to emphasize that we are thinking of ω as being equal to the set of all finite ordinals, i.e., as $\omega = \{0, 1, 2, 3, \cdots\}$, we write $[0, \omega)$ instead of just ω. This notation means just what it says: $[0, \omega)$ is the interval of all ordinal numbers that are greater than or equal to 0 and less than ω—it is the set of all finite ordinal numbers with their usual ordering.

The cardinal number associated with the ordinal number ω is then just the cardinality of the set $[0, \omega)$, so the proof of the following theorem is easy.

3.2. Theorem. $|\omega| = \aleph_0$.

The following exercises should now be easy, but understanding them is *crucial*.

3.3. Exercises.

1) There is no ordinal number immediately before ω, i.e., there is no ordinal number which when increased by 1 gives ω. In other words, $\omega - 1$ does not exist.
2) Why can't $\omega - 1$ be thought of as the set $\{0, 2, 3, 4, \cdots\}$?

Recall that for finite ordinal numbers (non-negative integers) n and m, we have $n < m$ as ordinal numbers if and only if $n < m$ as cardinal numbers (where $n < m$ as ordinal numbers means that $n \subsetneqq m$ when they are in canonical form). We can extend this to ω because it is true that if n is an ordinal number, then $n < \omega$ as ordinal numbers if and only if $n < \aleph_0$ as cardinal numbers. (This is just a fancy way of saying that an ordinal number is less than ω if and only if it is a finite ordinal number.) This might lead us to conjecture that given any two ordinal numbers a and b, then $a < b$ as ordinal numbers if and only if $a < b$ as cardinal numbers. Such a conjecture would be false! For example, consider the set

$$\{0, 1, 2, 3, \cdots, \omega\}$$

which consists of all finite ordinal numbers together with *one* new element, namely the number ω (remember how we went from 5 to 6; it is the same idea here. We are adjoining only *one* new element to the set $[0, \omega)$.) This set certainly looks like the canonical form of an ordinal number and it is certainly larger than ω because it contains $\omega = [0, \omega)$ as a proper subset. But its cardinality is $\aleph_0$ (why?). We denote this new set by the symbol $\omega + 1$, i.e.,

$$\omega + 1 = \{0, 1, 2, 3, \cdots, \omega\}.$$

Then we have $\omega < \omega + 1$ but $|\omega| = |\omega + 1|$, so we have an example of two different ordinal numbers whose corresponding cardinal numbers are the same—quite a departure from the case of finite ordinal numbers.

As a matter of fact, there are *infinitely many* ordinal numbers not equal to ω which have the same cardinality as ω. For example, we could write

$$\omega + 2 = \{0, 1, 2, 3, \cdots, \omega, \omega + 1\}$$

$$\omega + 3 = \{0, 1, 2, 3, \cdots, \omega, \omega + 1, \omega + 2\},$$

and so on, and it should not be too hard to imagine

$$\omega + \omega = \{0, 1, 2, 3, \cdots, \omega, \omega + 1, \omega + 2, \omega + 3, \cdots\},$$

which looks (set-wise) like two copies of $[0, \omega)$, one right after the other. Clearly $\omega + \omega$ is "twice as big" as ω when we think of them as ordinal numbers, but its cardinality is still countable (Why?). And the process goes on: to $\omega + \omega + 1$, to $\omega + \omega + \omega$, to $\omega + \omega + \omega + 1$—forever!

But—think back to when we first saw the number ω. We imagined that we had written down all of the finite ordinal numbers and then took ω to be the set consisting of all of these finite ordinal numbers. As such, ω was necessarily an infinite ordinal number—countably infinite of course: it was the first ordinal number that was not finite. We make the same kind of leap in cardinality here. Imagine that we have written down all of the ordinal numbers that are either finite or have cardinality $\aleph_0$, i.e., all of the countable ordinal numbers (remember that countable means either finite or countably infinite). The next ordinal number in line, then, would necessarily have to be an uncountable set (Why?). The smallest uncountable ordinal number (which, in canonical form, is the ordered set consisting of all the countable ordinal numbers) is usually denoted by the symbol Ω (sometimes it is called ω_1). The cardinality of this first uncountable ordinal is of course not $\aleph_0$ (if it were, Ω would be countable); instead, it should be clear that $|\Omega| > \aleph_0$.

The symbol used for the cardinality of Ω is $\aleph_1$, so $\aleph_1$ is then an infinite cardinal number which is greater than $\aleph_0$. Again we see that *not all infinite sets are the same size.*

The set A of all finite ordinal numbers has cardinality $\aleph_0$ and represents the number ω. The set B of all countable ordinal numbers has cardinality $\aleph_1$ and represents the number Ω. In other words, there are $\aleph_0$ finite ordinal numbers and $\aleph_1$ countable ordinal numbers. Also, since $\omega = [0, \omega) \subsetneqq [0, \Omega) = \Omega$, it follows that $\omega < \Omega$; but unlike the other ordinal numbers that we have seen which are greater than ω, Ω is uncountable, and $\aleph_0 < \aleph_1 (|\omega| < |\Omega|)$.

It is also important to note that the set C of countably infinite ordinal numbers has cardinality $\aleph_1$ so there are $\aleph_1$ ordinal numbers whose cardinality is $\aleph_0$.

4. ORDINAL NUMBERS GREATER THAN Ω

We have seen that there are $\aleph_1$ ordinal numbers whose cardinality is $\aleph_0$, namely the numbers

$$\omega, \omega + 1, \omega + 2, \omega + 3, \cdots, \omega + \omega, \cdots.$$

We can generalize this to obtain larger and larger ordinal numbers of larger and larger cardinality.

4.1. Definition.
   a) An **initial ordinal** is the smallest ordinal number with its cardinality.
   b) A **limit ordinal** is an ordinal number with no immediate predecessor.

Thus ω is the initial ordinal of cardinality $\aleph_0$, Ω is the initial ordinal of cardinality $\aleph_1$, and both are also limit ordinals. In fact, any initial ordinal is necessarily a limit ordinal (Why?), but there exist limit ordinals which are not initial ordinals (What are some?)

Summarizing the construction so far, we have the finite ordinal numbers $0, 1, 2, 3, \cdots, n, \cdots$ where $0 = \emptyset$ and $n = \{0, 1, 2, 3, \cdots, n - 1\}$ in canonical form. There are $\aleph_0$ such finite ordinal numbers. The smallest ordinal number that is not finite is denoted by ω (or ω_0) and is, in canonical form,

$$\omega = \{0, 1, 2, 3, \cdots, n, \cdots : n \in \mathbf{Z}^+\}.$$

It is the initial ordinal number of cardinality $\aleph_0$. The countably infinite ordinal numbers (i.e., those whose cardinality is $\aleph_0$) are then

$$\omega_0, \omega_0 + 1, \omega_0 + 2, \omega_0 + 3, \cdots, \omega_0 + \omega_0, \cdots$$

and there are $\aleph_1$ such countable ordinal numbers. The smallest ordinal number whose cardinality is not countable is denoted by Ω (or ω_1) and, in canonical form, is equal to the set of all finite and countably infinite ordinal numbers ordered in the usual way. The ordinal number ω_1 is the initial ordinal of cardinality $\aleph_1$.

To generalize, the ordinal numbers with cardinality $\aleph_1$ are

$$\omega_1, \omega_1 + 1, \omega_1 + 2, \cdots, \omega_1 + \omega_0, \omega_1 + \omega_0 + 1, \cdots,$$
$$\omega_1 + \omega_1, \omega_1 + \omega_1 + 1, \cdots.$$

The smallest ordinal number whose cardinality is greater than $\aleph_1$ is denoted by ω_2 (you can see the reason now for the alternate notation ω_0 for ω and ω_1 for Ω), and the cardinality of ω_2 is denoted by $\aleph_2$. Then there are $\aleph_2$ ordinal numbers whose cardinality is $\aleph_1$, and ω_2 is the initial ordinal number with cardinality $\aleph_2$. Then there are $\aleph_3$ ordinal numbers with cardinality $\aleph_2$ and the initial ordinal of cardinality $\aleph_3$ is denoted ω_3. Clearly this process can be continued, to $\aleph_4$ and ω_4, to $\aleph_{\omega_0}$ and ω_{ω_0}, to $\aleph_{\omega_1}$ and ω_{ω_1}, forever.

Thus we get an infinite collection of infinite initial ordinals

$$\omega = \omega_0 < \Omega = \omega_1 < \omega_2 < \cdots < \omega_{\omega_0} < \omega_{\omega_0+1} < \cdots < \omega_{\omega_1} < \cdots$$

and the corresponding infinite collection of their cardinal numbers (called *alephs*)

$$\aleph_0 < \aleph_1 < \aleph_2 < \cdots < \aleph_{\omega_0} < \aleph_{\omega_0+1} < \cdots < \aleph_{\omega_1} < \cdots.$$

It is easier to keep things straight if you observe that the subscript on the omegas and the alephs is the type of ordered set formed by the (transfinite) initial ordinals that precede the omega or aleph in question. For example, $\omega = \omega_0$ has no infinite initial ordinals before it, $\Omega = \omega_1$ has one (namely ω_0), ω_2 has 2, and so on.

Before concentrating more on the set of countable ordinals (which will be an important example in our study of topology), there is a question that should be asked. Recall that c, the cardinality of the real line, is uncountable. The question is: is c one of the alephs (i.e., is c the cardinality of some initial ordinal), and if so, which one is it? It can be proved that the answer to the first part is yes—it turns out that every infinite number is one of the alephs*; the second part is more subtle. Is $c = \aleph_1$? It should be clear that given that c is an aleph, then $c = \aleph_1$ if and only if the continuum hypothesis holds.

---

* You can find a proof and some good discussion in Sierpinski [13].

5. THE COUNTABLE ORDINALS

Given any finite set S consisting of non-negative integers, you can show that there exists a non-negative integer N such that $N \geq s$ for every element $s \in S$. This can be stated in a somewhat fancier way by saying that every finite subset of $[0, \omega)$ is bounded above, and generalizes to the following important theorem.

5.1. Theorem. Any countable subset of $[0, \Omega)$ is bounded above, i.e., if $S \subset [0, \Omega)$ with $|S| \leq \aleph_0$, then there exists a number a in $[0, \Omega)$ such that $a \geq s$ for every $s \in S$.
Hint for Proof: Suppose that $S = \{x_1, x_2, x_3, \cdots\}$ is a countable subset of $[0, \Omega)$ which is not bounded above. Show (using Theorem 1.7) that this forces $|[0, \Omega)| = \aleph_0$, a contradiction.

6. WELL-ORDERING AND THE PRINCIPLE OF TRANSFINITE INDUCTION

A very important fundamental property of ordinal numbers is that any collection of ordinal numbers is **well-ordered,** which means that any collection of ordinal numbers contains a smallest element. We cannot prove this because we do not have a definition of an ordinal number, but we will have need for it later, so we state it here as a fact for future reference. (You can see a rigorous justification in Wilder [18].)

6.1. Fact. Any set of ordinal numbers is well-ordered. In other words, any collection of ordinal numbers contains a smallest element.

6.2. Exercises.
1) Show that **R** is not well-ordered by exhibiting a subset of **R** which does not contain a smallest element.
2) Show that the set of positive real numbers is not well-ordered by exhibiting a collection of positive real numbers which does not contain a smallest element.

Mathematical induction is a procedure for proving results about the non-negative integers, and is based on the following fact. (A proof is not asked for.)

6.3. The Principle of Mathematical Induction. Let N be a non-negative integer and let $P(N)$ be a statement about N. Then
1) If $P(0)$ is true, and
2) If $n \in \mathbf{Z}^+$ and if the fact that $P(k)$ is true for all $k < n$ implies that $P(n)$ is also true,
    then $P(n)$ is true for all non-negative integers n.

6.4. Exercise.

Prove that for any non-negative integer n,

$$1 + 2 + 3 + \cdots + n = n(n + 1)/2.$$

The non-negative integers form a well-ordered set (indeed, the set of non-negative integers is the ordinal number $\omega = [0, \omega)$), and the principle of mathematical induction is in fact equivalent to the fact that the non-negative integers are well-ordered, as you can show in the following theorem.

6.5. Theorem. The fact that the set of non-negative integers is well-ordered implies that the principle of mathematical induction holds for the set of non-negative integers. Conversely, the fact that the principle of mathematical induction holds for the non-negative integers implies that the set of non-negative integers is well-ordered.

In view of Theorem 6.5, it should not be surprising that a form of induction holds for any well-ordered set. In particular, if α is an ordinal number, then $[0, \alpha)$ is well-ordered. Use this fact to prove the following.

6.6. Theorem. (The principle of transfinite induction.) Let α be an ordinal number, and let $P(x)$ be a statement about ordinal numbers less than α. Then

1) If $P(0)$ is true, and
2) If $\beta < \alpha$ and $P(\gamma)$ true for all $\gamma < \beta$ implies that $P(\beta)$ is true, then $P(x)$ is true for all ordinal numbers $x < \alpha$.

chapter three

Some Familiar Topological Spaces and Basic Topological Concepts

A **topology** on a set is a collection of subsets of that set which satisfies certain properties (which we will give later). The sets that belong to the topology are called **open sets,** and their complements are **closed sets.** When we have a topology $\mathfrak{I}$ defined on a set X, the pair $(X, \mathfrak{I})$ is called a **topological space.** In this chapter we will look at some very familiar sets which happen to be topological spaces, and some familiar ideas which are topological in nature. You may be surprised at how much topology you already know.

A word about content and rigor is in order before we start. All of the theorems and some of the exercises in this chapter require a completely rigorous, formal proof. Some of the exercises, though, involve material which is usually considered too "advanced" for the background that we have now, and indeed we do not have the necessary background to approach this material from a completely rigorous point of view. But these "advanced topics" are fascinating and well worth seeing now, if only to get an idea of things to come. As a result of the desire to discuss this material before we are totally ready for it, some of the exercises in this chapter begin with "convince yourself" rather than "prove." Although there can be no doubt about whether a proof is a proof or not, there may well be some subjectivity as to whether you have in fact convinced yourself of something. What we really mean by "convince yourself" is to be sure that you understand what is going on, *and* that you can explain it, at least intuitively, to someone with roughly the same background that you have. Usually a carefully drawn picture and a well thought out explanation of it is enough to "convince yourself."

1. THE TOPOLOGY ON THE REAL LINE

To define a topology on the real line we need to decide what subsets of the line will be open sets. Certainly it is reasonable to want ordinary open intervals to be open sets, and so they should belong to the topology (remember that a topology is a collection of sets). When you think about it though, the "open-ness" of an open interval does not really have anything to do with its being an interval that is all in one piece; rather, an open interval is open because there is no point *in* it where it suddenly "ends." For any point in the open interval, all of the points both a little above and a little below the point we are looking at also belong to the open interval. An open interval sort of "fades out" at its end points, and this is the property that makes it open. (Compare this with closed intervals.) This property of open intervals fading out at their end points can be stated precisely by saying that for every point x in an open interval I, there is a positive number r such that the particular open interval $(x - r, x + r) \subset I$. We will use this property to define open subsets of the line in general.

1.1. Definition. A subset $A \subset \mathbf{R}$ is **open** if for every point $x \in A$, there is a positive number r such that $(x - r, x + r) \subset I$.

Thus a subset A of $\mathbf{R}$ is open if and only if for each point $x \in A$, there is a number $r > 0$ such that A contains all points of $\mathbf{R}$ whose distance from x is less than r. This makes precise the idea of an open set containing all points both a little bit above and a little bit below any point in it—a little bit above or a little bit below means less than r above or below.

1.2. Exercises.
1) Let $A = (0, 1) \cup (1, 3)$. For the given $x \in A$, give a value of $r > 0$ such that $(x - r, x + r) \subset A$.
   a) $x = \frac{3}{4}$.
   b) $x = 2$.
   c) $x = \frac{9}{8}$.
2) Prove that $A = (0, 1) \cup (1, 3)$ is an open subset of $\mathbf{R}$.
3) Prove that an ordinary open interval is an open subset of $\mathbf{R}$ but that an open set need *not* be an open interval.
4) State precisely what it means when a subset A of $\mathbf{R}$ is *not* open.
5) Prove that the following subsets of $\mathbf{R}$ are not open.
   a) The set of rational numbers.
   b) A set consisting of a single point.
   c) An interval of the form $[a, b)$, where $a < b$.
   d) The set $A = \{x \in \mathbf{R} : x \neq 1/n, \text{ for } n \in \mathbf{Z}^+\}$.

The key idea involved in the discussion of open subsets of $\mathbf{R}$ is that of *distance*. As you know from calculus, the distance between two points x_1

and x_2 in **R** is given by $|x_1 - x_2|$, so an interval $(x - r, x + r)$ can be written as

$$(x - r, x + r) = \{y \in \mathbf{R}: |x - y| < r\}.$$

We call this set an **r-ball** centered at the point x (for reasons that will be clear later), and denote it by the symbol $S_r(x)$. Thus, an r-ball around a point $x \in \mathbf{R}$ is the set of points of **R** whose distance from x is less than r:

$$S_r(x) = (x - r, x + r) = \{y \in \mathbf{R}: |x - y| < r\}.$$

The definition of an open subset of **R** could then be rewritten using this new terminology as

A subset $A \subset \mathbf{R}$ is **open** if for any point $x \in A$, there exists a positive number r such that $S_r(x) \subset A$.

The following theorem is important because it gives the conditions that we will use to define topological spaces in general.

1.3. Theorem.
  a) The union of any collection of open subsets of the real line is also an open subset of the line.
  b) The intersection of any finite collection of open subsets of the real line is also an open subset of the line.
  c) Both the empty set and **R** itself are open subsets of the real line.

1.4. Exercise.
  Give an example of an infinite collection of open subsets of the real line whose intersection is not open, thus showing that the finiteness condition in Theorem 1.3(b) is necessary.

1.5. Definition. A subset F of **R** is **closed** if its complement $(\mathbf{R} - F)$ is open.

It is immediate from this definition that **R** and $\emptyset$ are both closed. But they are both open, too, as we saw in Theorem 1.3. Thus *a set can be both open and closed.* Also, a set may be neither open nor closed (give an example). Thus *the fact that a set is not open does not necessarily mean that it is closed and the fact that a set is not closed does not necessarily mean that it is open.*

Since closed sets are the "opposites" of open sets, they satisfy a theorem which is just the opposite of Theorem 1.3 for open sets. Use the De Morgan laws and Theorem 1.3 to prove it.

1.6. Theorem.
  a) The intersection of any collection of closed sets is closed.
  b) The union of any finite collection of closed sets is closed.
  c) $\emptyset$ and **R** itself are both closed.

1.7. Exercises.

1) State precisely what it means when a subset of **R** is not closed. (Do this in terms of points; saying that a set is not closed if its complement is not open is true, but is not what we want here.)

2) Which of the following subsets of **R** are closed? Which are open?
   a) The set **Z** of integers.
   b) The set of rational numbers.
   c) A set consisting of a single point.
   d) An interval of the form $[a, b)$, where $a < b$.
   e) The set $A = \{x \in \mathbf{R} : x \neq 1/n \text{ for } n \in \mathbf{Z}^+\}$.
   f) The set $A = \{x \in \mathbf{R} : x \neq 1/n \text{ for } n \in \mathbf{Z}^+ \text{ and } x \neq 0\}$.

3) Prove that an ordinary closed interval is a closed subset of **R**, but a closed set need not be a closed interval.

4) Give an example of an infinite collection of closed subsets of **R** whose union is not closed, thus showing that the finiteness condition in Theorem 1.6(b) is necessary.

2. THE EUCLIDEAN PLANE E² AS A TOPOLOGICAL SPACE

We defined what it means for a subset of **R** to be open by saying that it must contain an r-ball around each of its points (where r usually depends on the particular point that we are looking at). This same idea of what "open" means should make sense whenever the notion of what an r-ball is makes sense. It should be clear that in the definition of r-ball, the important thing is being able to measure the distance between any two points, and what kind of points they are makes no difference at all as long as we can measure the distance between them.

Surprisingly, it is not always possible to measure distance in a meaningful way, as we will see later. It *is* possible when we are working with the real line, as we have seen: if $x, y \in \mathbf{R}$, then the distance between x and y is given by $|x - y|$. Letting $d(x, y)$ denote the distance between the points x and y, we have then for $x, y \in \mathbf{R}$ that

$$d(x, y) = |x - y|.$$

Then we say that a subset A of **R** is open if for each point $x \in A$, there is a positive number r such that the r-ball around x

$$S_r(x) = \{y \in \mathbf{R} : |x - y| < r\} = \{y \in \mathbf{R} : d(x, y) < r\}$$

is totally contained in A.

It is also possible to measure distance in a meaningful way between points in the Euclidean plane. You may recall the "distance formula" from elementary calculus: for two points (x_1, x_2) and (y_1, y_2) in the plane, the distance between (x_1, x_2) and (y_1, y_2) is given by

$$d((x_1, x_2), (y_1, y_2)) = \sqrt{(x_1 - y_1)^2 + (x_2 - y_2)^2}.$$

Using this distance formula, we can talk about r-balls around points in the plane: in the plane, an r-ball is still the set of all points whose distance from a fixed point is less than r, only now it looks like a disk (without its edge) instead of like an open interval. Then we can say that a subset A of the plane is open if for each point $(x_1, x_2) \in A$, there is an $r > 0$ such that the r-ball $S_r(x_1, x_2)$, consisting of all points in the plane whose distance from (x_1, x_2) is less than r, is totally contained in A.

Thus in both the line and the plane, a set is open if it contains an r-ball around each of its points. Each of these two spaces—the line and the plane—is a special case of something called a *metric space*, and the two distance formulas

$$d(x, y) = |x - y| \text{ on the line, and}$$

$$d(x, y) = \sqrt{(x_1 - y_1)^2 + (x_2 - y_2)^2},$$

where $x = (x_1, x_2)$ and $y = (y_1, y_2)$ in the plane,

are special cases of what is called a *metric*. A metric on a set is a function that assigns a non-negative real number to each pair of elements in the set, and this number is called the *distance* between the pair of elements. A function that measures the distance between points—a metric—must satisfy certain properties which, when you think about it, a measure of distance *should* satisfy. (It is usually the case in mathematics that things are as they ought to be; a lot of confusion can be alleviated if instead of memorizing abstract definitions, you try to understand why the definition ought to be what it is.) The properties that a measure of distance should satisfy are:

1) The distance between any two points is never negative.
2) The distance between distinct points is positive, and the distance from a point to itself is 0.
3) The distance between two points does not depend on the direction in which we measure it; i.e., the distance from P_1 to P_2 is the same as the distance from P_2 to P_1.
4) The distance from one point to another when measured directly is never more than the distance between these points measured by going through some particular third point. ("The shortest distance between two points is a straight line"; "the hypotenuse in a right triangle is shorter than the sum of the lengths of the other two sides.") This property, called the *triangle inequality*, is satisfied by distance functions even in non-Euclidean geometries—except that the meaning of "straight line" may have to be changed.

The following definition makes all of this precise. When you read it don't confuse the pair (x, y) of points of X with the single point (x, y) in the plane. In this definition, $d(x, y)$ means the distance between the two points x and y of X (as it did before), and what x and y themselves look like depends on the particular set X that they come from. For example, if X is the plane, then $d(x, y)$ means $d((x_1, x_2), (y_1, y_2))$, where $x = (x_1, x_2)$ and $y = (y_1, y_2)$.

2.1. Definition. Let X be a set. A function

$$d: X \times X = \{(x, y): x, y \in X\} \to \mathbf{R}$$

is a **metric** on X if it satisfies the following:
1) $d(x, y) \geq 0$ for all $x, y \in X$.
2) $d(x, y) = 0$ if and only if $x = y$.
3) $d(x, y) = d(y, x)$ for all $x, y \in X$.
4) $d(x, y) \leq d(x, z) + d(y, z)$ for all $x, y, z \in X$. (This property of d is called the **triangle inequality.**)

When d is a metric on a set X, then an **r-ball** centered at $x \in X$ is the set

$$S_r(x) = \{y \in X: d(x, y) < r\},$$

and is the set of all points of X whose distance from x is less than r, where r is a positive real number.

Continuing to follow the example of the line and the plane, we define the *metric topology on X induced by a metric d* as follows.

2.2. Definition. Let d be a metric on the set X. A subset A of X is open in the **metric topology on X induced by d** if, for every point $x \in A$, there is an $r > 0$ such that the r-ball centered at x, $S_r(x)$, is contained in A. The set X together with the topology induced by a metric d on X is called a **metric space,** and is usually written as (X, d).

Thus a subset of a metric space is open in the metric topology if and only if it contains an r-ball around each of its points.

2.3. Exercises.
1) Show that the absolute value formula, $d(x, y) = |x - y|$, is indeed a metric on the real line. Describe the **1-ball** centered at 0 in the topology induced by this metric.
2) Show that the distance formula

$$d(x, y) = \sqrt{(x_1 - y_1)^2 + (x_2 - y_2)^2},$$

where $x = (x_1, x_2)$, $y = (y_1, y_2)$, is a metric on the Euclidean plane. Describe the 1-ball centered at $(0, 0)$ in the topology induced by this metric.

3) Define the obvious metric for Euclidean 3-space, $E^3 = \mathbf{R} \times \mathbf{R} \times \mathbf{R} = \{(x, y, z) : x, y, z \in \mathbf{R}\}$. Describe the 1-ball centered at the point $(0, 0, 0)$ in the topology induced by this metric. (In this topology, an r ball really is a ball—hence the name "r-ball.")

4) Show that for any metric space (X, d),
    a) The union of any collection of open sets is open.
    b) The intersection of any finite collection of open sets is open.
    c) The empty set and X itself are open.

5) A set can have more than one metric defined on it, and different metrics may give rise to different topologies.
    a) Let $X = \mathbf{R}$ and define a metric on X by $d(x, y) = 1$ if $x \neq y$, $d(x, y) = 0$ if $x = y$. Prove that d is a metric on X. What is $S_{1/2}(0)$ in this metric? Is (X, d) the same space as $\mathbf{R}$ with its usual metric topology? In other words, does this metric give rise to the same topology on $\mathbf{R}$ as the usual metric does?
    b) Let X be the Euclidean plane and define a metric on X by $d((x_1, x_2), (y_1, y_2)) = |x_1 - y_1| + |x_2 - y_2|$. Prove that d is a metric on the plane, and describe the r-balls in this metric. Does this metric give rise to the same topology as the usual metric on the plane?

Since a set can have more than one metric defined on it and since different metrics may give rise to different topologies, we will refer to the absolute value metric on $\mathbf{R}$ and the distance formula metric on the plane as the usual metrics for these sets.

3. SOME FAMILIAR SUBSETS OF R AS TOPOLOGICAL SPACES

Any subset of $\mathbf{R}$ can be made into a metric space by using the usual metric on $\mathbf{R}$. Thus, if $S \subset \mathbf{R}$, the distance between points x and y in S is just $d(x, y) = |x - y|$. Using this metric, S can be made into a metric space.

3.1. Definition. Let $S \subset \mathbf{R}$. A subset A of S is open in the **metric topology on S induced by the absolute value metric d** if for any point $x \in A$, there exists an $r > 0$ such that $S_r(x) \cap S \subset A$.

Note that we do *not* require that the entire r-ball $S_r(x)$ be contained in A in order that A be open in (S, d). Rather, only the points of $S_r(x)$ that are also points of S need belong to A. For example, if $S = \mathbf{Q}$, the set of rational numbers, then the interval consisting of only rational numbers which are greater than 1 and less than 2 *is* open in the metric space $(\mathbf{Q}, d)$, even though it is *not* open in $\mathbf{R}$. Thus a set may be open in a subset of $\mathbf{R}$ with the metric topology while it is not open in $\mathbf{R}$ itself, so intuition about what "open" means may have to be used with care.

As a second example, when $S = [1, 2] \subset \mathbf{R}$ is given the metric topology induced by the absolute value metric, sets of the form $[1, b)$ and $(a, 2]$ are open in the resulting metric space, even though they may not be open in $\mathbf{R}$. Again, it is important to realize that only the points of S matter when we are talking about open subsets of S.

Finally, let $S = \mathbf{Z}$, the set of integers. In the metric space that results when $\mathbf{Z}$ is given the topology induced by the absolute value metric, every point is an open set, as you are asked to show in the exercises below. The resulting space is an example of what is called a *discrete* space.

3.2. Definition. A topological space in which every set consisting of a single point is open is called a **discrete space.**

3.3. Exercises.

1) Let $(\mathbf{Q}, d)$ be the space of rational numbers with the metric topology induced by the absolute value metric d.
   a) Show that the set $\{y \in \mathbf{Q} : 1 < y < 2\}$ is open in $(\mathbf{Q}, d)$.
   b) Show that the set $[\sqrt{2}, \sqrt{3}] \cap \mathbf{Q}$ is open in $(\mathbf{Q}, d)$.
2) Let $S = [1, 2]$ be given the topology induced by the absolute value metric. Show that $[1, \frac{1}{2})$, $(\frac{1}{2}, 2]$ and $[1, 2]$ are all open in (S, d).
3) Let $(\mathbf{Z}, d)$ be the space of integers with the metric topology induced by the absolute value metric. Show that *every* subset of $\mathbf{Z}$ is open in $(\mathbf{Z}, d)$. In particular, any set consisting of a single point is open in $(\mathbf{Z}, d)$.
4) Let (X, d) be a discrete space. Show that every subset of X is open in (X, d).

4. THE ORDER TOPOLOGY

We defined the topology on the real line by using its metric, $d(x, y) = |x - y|$, and saying that a subset A of $\mathbf{R}$ is open if for each point $x \in A$, there is an $r > 0$ such that the interval $(x - r, x + r) \subset A$. Another way to look at this topology is to avoid the metric entirely and use arbitary open intervals instead, which the following theorem justifies.

4.1. Theorem. A subset A of $\mathbf{R}$ is open in the metric topology induced by the absolute value metric if and only if for each point $x \in A$, there exist real numbers a and b with $a < x < b$, such that $x \in (a, b) \subset A$.

The topology that we get on $\mathbf{R}$ by using the definition of open set given in Theorem 4.1 is called the *order topology* on $\mathbf{R}$. The theorem shows that the order topology on $\mathbf{R}$ and the metric topology on $\mathbf{R}$ induced by the absolute value metric are the same: a set is open in one if and only if it is open in the other. The advantage of looking at the topology on $\mathbf{R}$ in this new way is that the metric on $\mathbf{R}$ need not be mentioned at all; the really important thing from this new point of view is that open intervals exist, because "open set" is defined in terms of open intervals. It should make sense that whenever open intervals make sense in a set, then we can define a topology on that set as we just did for **R**, in terms of these open intervals (whether the set has a metric defined on it or not). To find out when open intervals make enough sense to be useful, we have the following definition.

4.2. Definition. Let X be a set. The relation $\leq$ is a **total order** relation on X if it satisfies the following:
1) For any $x \in X$, $x \leq x$ ($\leq$ is *reflexive*).
2) If for $x, y \in X$, both $x \leq y$ and $y \leq x$, then $x = y$ ($\leq$ is *anti-symmetric*).
3) If for $x, y, z \in X$, both $x \leq y$ and $y \leq z$, then $x \leq z$ ($\leq$ is *transitive*).
4) If $x, y \in X$, then either
   a) $x \leq y$ and $x \neq y$ (written $x < y$),
   or
   b) $x = y$,
   or
   c) $y \leq x$ and $x \neq y$ (written $y < x$ or $x > y$).

4.3. Exercise.

Convince yourself that the real line is a totally ordered set. Give an example of a set with an order relation that satisfies (1), (2), and (3) of Definition 4.2, but does *not* satisfy (4). Such an ordered set is called **partially ordered** since not every pair of elements can be compared. [*Hint:* Consider $\mathcal{P}(X)$ with "$\leq$" thought of as "subset of"].

The order topology on a totally ordered set is defined as follows.

4.4. Definition. Let $(X, \leq)$ be a totally ordered set (i.e., $\leq$ is a total order relation on the set X). Then $U \subset X$ is open in the **order topology** on $(X, \leq)$ if and only if for each point $x \in U$, one of the following holds:
1) x is the first point of X and there exists a point $b \in X$ such that the interval $[x, b) = \{p \in X : x \leq p < b\}$ is contained in U, or
2) x is the last point of X and there exists a point $a \in X$ such that the interval $(a, x] = \{p \in X : a < p \leq x\}$ is contained in U, or
3) x is neither the first nor the last point of X and there exist points a and b in X such that $x \in (a, b) = \{p \in X : a < p < b\}$, and $(a, b) \subset U$.

Thus a set U is open in the order topology on a totally ordered set X if and only if each point of U is contained in an open interval which in turn is totally contained in U (with appropriate modifications in case U contains the first or last point of X, if there is one).

The rest of this section requires a knowledge of ordinal numbers.

By Theorem 4.1, the topology that we defined on **R** is the order topology on **R**. Another important totally ordered set is [0, Ω), the set of countable ordinals, and when we give this set the order topology as defined above, we get an important topological space: the **space of countable ordinals.**

Even though **R** and Ω (= [0, Ω)) have both been given the order topology, they are very different looking topological spaces, as the following exercises illustrate.

4.5. Exercises.
   1) Every open subset of **R** has cardinality c.
   2) For any positive integer n, there are open subsets of [0, Ω) with cardinality n. In particular, there are points x in [0, Ω) such that the singleton set $\{x\}$ is an open set. But [0, Ω) is not discrete—not every singleton set is open.
   3) There are open subsets of [0, Ω) of cardinality $\aleph_0$ and there are open subsets of [0, Ω) of cardinality $\aleph_1$.

Thus there are finite, countably infinite, and uncountably infinite subsets of the ordered space [0, Ω) which are open, while every open subset of the ordered space **R** is uncountably infinite. Besides these differences, another important difference between the two ordered topological spaces **R** and [0, Ω) involves their respective "infinities" (∞ and Ω). If we imagine for the moment that $\infty \in$ **R** as its largest point, and if we look at [0, Ω] as [0, Ω) with the point Ω included, we can consider (order topology) open sets containing ∞ in **R** $\cup \{\infty\}$, and containing Ω in [0, Ω]. The difference is that there is a countable collection of open sets of **R** $\cup \{\infty\}$ whose intersection is the point ∞ (try the collection $\mathcal{C} = \{(n, \infty] : n \in \mathbf{Z}^+\}$; show that each member of $\mathcal{C}$ is open in the order topology on **R** $\cup \{\infty\}$ (even though they might not look like it) and show that $\cap\mathcal{C} = \{\infty\}$. This is just another way of saying that $\lim_{n\to\infty} n = \infty$.) On the other hand, *no* countable collection $\mathcal{C}$ of open subsets of [0, Ω] can have $\cap\mathcal{C} = \{\Omega\}$ (Why not?), so for any set $\{a_n : n \in \mathbf{Z}^+\}$ of countable ordinal numbers, $\lim_{n\to\infty} a_n < \Omega$. This may not seem like much, but it is really a very important difference between the two spaces. (While we are on the subject, you should show that Ω is the only point of [0, Ω] with this property—for every other point x of [0, Ω] there *is* a countable collection $\mathcal{C}$ of open sets with $\cap\mathcal{C} = \{x\}$.)

Another important set which is totally ordered and can therefore be given the order topology is the set of finite ordinal numbers, $\mathbf{Z}^+ \cup \{0\} = \omega =$ [0, ω). The resulting topological space is the same as that obtained when $\mathbf{Z}^+ \cup \{0\}$ is given the metric topology induced by the absolute value metric (Problem 1 in the exercises below). Thus ω, when given the order topology, is the discrete space of finite ordinal numbers. (Recall that neither **R** nor [0, Ω) is discrete; recall also that [0, Ω) is "more discrete" than **R** is—*some* of its points are open.)

4.6. Exercises.

1) Show that $\omega = [0, \omega)$ with the order topology is a discrete space, so is the same as $\mathbf{Z}^+ \cup \{0\}$ when given the metric topology induced by the absolute value metric on $\mathbf{R}$.

2) The definition of the order topology is given in terms of open intervals. Let $(X, \leq)$ be a totally ordered set and give it the order topology. Show that a subset of X is open in the order topology if and only if it is a union of open intervals. Because of this, we say that the collection of open intervals is a *basis* for the order topology on a totally ordered set.

5. SEQUENCES IN METRIC SPACES

The idea of convergence of a sequence is one of the most important ones in analysis and topology. Unfortunately, it is a rather subtle idea which some students find hard to understand at first. With a little study and patience you can master it, and once you do, it will be one of the simplest and most important tools at your disposal.

5.1. Definition. A **sequence** in a set X is a function that can be written as a function from the positive integers into X.

Instead of using the ordinary function notation $S(n)$ for the value of the sequence (function) S at the integer n, we usually pick some appropriate letter, say x, and denote the value of the sequence at n by x_n. (Of course other letters can be, and often are, used.) Actually, we usually don't think of a sequence as a function at all; rather, we regard a sequence in X as a countable set of points of X which are indexed and *given an ordering* by the positive integers. The ordering is simply that if $n > m$, then x_n is further out in the sequence than x_m, regardless of how the points x_n and x_m are related to each other in X. For example, consider the sequence in $\mathbf{R}$ defined by $S(n) = x_n = 1/n$, for $n = 1, 2, 3, \cdots$. Written out, this sequence looks like

$$1, \tfrac{1}{2}, \tfrac{1}{3}, \tfrac{1}{4}, \cdots.$$

Notice that even though $\tfrac{1}{4} < \tfrac{1}{2}$, $\tfrac{1}{4}$ is further out in the sequence than $\tfrac{1}{2}$, because $4 > 2$.

As another example, consider the sequence defined by $x_n = 1$ if n is even and $x_n = -1$ if n is odd (where $n \in \mathbf{Z}^+$). This sequence simply alternates between -1 and 1 and looks like

$$-1, 1, -1, 1, \cdots.$$

In this sequence, for example, $x_{20} = x_{10}$ as points in $\mathbf{R}$, but x_{20} is still further out in the sequence than x_{10}.

We will usually denote a sequence of points x_1, x_2, x_3, $\cdots$ in X by $\{x_n : n \in \mathbf{Z}^+\}$ or, sometimes, for convenience, as $\langle x_n \rangle$. It is important to be aware of the fact that this notation means more than just a set of points; it denotes these points with the ordering that they get from the positive integers. For example, the sequence $\{x_n : n \in \mathbf{Z}^+\}$ defined by $x_n = 0$ for all $n \in \mathbf{Z}^+$ is *not* just the set consisting of the single element 0; $\{0\}$ might be called the *point-set determined by the sequence* but it is not the sequence. Rather, the sequence consists of countably many 0's, all different in the sense that each corresponds to a different positive integer. (This distinction between the sequence and the point-set determined by the sequence is the same as considering a function as being different from the graph of the function; it is very important to make the distinction with sequences, as we will see.)

The idea of *convergence* is based on the ordering of a sequence induced by the positive integers. Intuitively, we say that a sequence $\{x_n : n \in \mathbf{Z}^+\}$ *converges* to a point x if the terms of the sequence eventually get close to x and stay close as n gets larger and larger. We make this precise as follows.

5.2. Definition. Let $\{x_n : n \in \mathbf{Z}^+\}$ be a sequence of points of a metric space (X, d), and let $x \in X$. Then $\{x_n : n \in \mathbf{Z}^+\}$ **converges** to x (written $x_n \rightarrow x$ or $\lim_{n \to \infty} x_n = x$) if for any positive real number r (no matter how small) there exists a positive integer N such that $x_n \in S_r(x)$ for all $n \geq N$.

The condition "there exists an integer N such that $x_n \in S_r(x)$ for all $n \geq N$" is used a lot, and for convenience, we will say that when a sequence satisfies this property, then it is *ultimately* in $S_r(x)$.

Thus we can restate the definition of convergence as follows:

A sequence $\{x_n : n \in \mathbf{Z}^+\} \subset (X, d)$ **converges** to the point $x \in X$ (written $x_n \rightarrow x$ or $\lim_{n \to \infty} x_n = x$) if for any real number r (no matter how small) x_n is ultimately in $S_r(x)$.

Thus the idea that a sequence in a metric space converges to a point if it gets close to the point and stays close, as n gets larger and larger, is very clear: in order to converge to a point, a sequence must eventually get within a distance less than r of the point and stay at least that close to the point as n gets larger and larger, no matter how small the distance r is.

5.3. Exercises.
1) State precisely what it means when a sequence in (X, d) does not converge to the point $x \in X$.
2) Let $\{x_n : n \in \mathbf{Z}^+\}$ be the sequence in $\mathbf{R}$ defined by $x_n = 1/n$. Does this sequence converge? To what? Prove it.
3) Let $\{x_n : n \in \mathbf{Z}^+\}$ be the sequence in $\mathbf{R}$ defined by $x_n = (-1)^n(1/n)$. Does this sequence converge? To what? Prove it.

4) Let $\{x_n : n \in \mathbf{Z}^+\}$ be the sequence in $\mathbf{R}$ defined by $x_n = (-1)^n$. Does this sequence converge? To what? Prove it.

The following theorem gives an important topological property of metric spaces.

5.4. Theorem. A sequence in a metric space can converge to at most one point.
[*Hint for proof:* Think about the real line.]

The property that a sequence in a space can converge to at most one point is sometimes called **uniqueness of convergence.** Surprisingly, not every space satisfies this property, and we will see some examples in the next chapter. We should also point out that, even in a metric space, a sequence need not converge at all (give an example). Theorem 5.4 says that *if* a sequence in a metric space converges, then it converges to exactly one point; it does *not* say that every sequence in a metric space must converge.

Recall that a subset of a topological space is closed if and only if its complement is open, i.e., if and only if its complement belongs to the topology. The following theorem relates sequences to the idea of a set being closed in a metric space.

5.5. Theorem. A subset S of a metric space X is closed if and only if whenever a sequence of points of S converges to a point $x \in X$, then $x \in S$.
[*Hint for proof:* Think about the real line. A closed set need *not* be a closed interval, even in $\mathbf{R}$!]

We might say that if there is a sequence inside a set S that converges to a point x, then x is *close* to S (how else could the sequence get close to x?) Thus in a metric space (and, in particular, in the real line) Theorem 5.5 says that a set is closed if and only if it contains all points of the space that are close to it.

5.6. Exercises.
  1) Use Theorem 5.5 to decide if the following subsets of the real line are closed (in the metric topology induced by the absolute value metric).
      a) $[0, 1]$
      b) $[1, \infty)$
      c) $\{x \in \mathbf{R} : x = 1/n \text{ for } n \in \mathbf{Z}^+\}$
      d) $\{x \in \mathbf{R} : x = 1/n \text{ for } n \in \mathbf{Z}^+, \text{ or } x = 0\}$
      e) $\{x \in \mathbf{R} : x = 1/\sqrt{n} \text{ for } n \in \mathbf{Z}^+\}$
  2) Is $(\mathbf{R} - \mathbf{Q}) \cap \{x \in \mathbf{R} : x = 1/\sqrt{n} \text{ for } n \in \mathbf{Z}^+\}$ a closed subset of the *irrationals* with the metric topology induced by the absolute value metric?

3) Let $\{p_n = (x_n, y_n) : n \in \mathbf{Z}^+\}$ be a sequence in E^2, the Euclidean plane. Prove that $p_n \to p = (x, y)$ if and only if both $x_n \to x$ and $y_n \to y$. (Thus a sequence in the plane converges if and only if it converges "coordinate-wise.")

4) Prove that every real number can be written as the limit of a convergent sequence of *rational* numbers, i.e., if $r \in \mathbf{R}$, exhibit a sequence $\{x_n : n \in \mathbf{Z}^+\} \subset \mathbf{Q}$ such that $x_n \to r$.

5) A real number r is called the **least upper bound** or **supremum** of a set $S \subset \mathbf{R}$ if
   a) $r \geq s$ for all $s \in S$, and
   b) If $t \in \mathbf{R}$ is a real number such that $t \geq s$ for all $s \in S$, then $t \geq r$.
   The least upper bound of a set S is denoted by $supS$. Prove that if $S \subset R$ and supS exists, then there exists a sequence of points of S that converges to supS.

6) Prove that if $S \subset \mathbf{R}$ and supS exists, then if S is closed, supS $\in S$. Is the converse true?

We have said that a sequence in (X, d) converges to a point x if it is ultimately in $S_r(x)$ for any $r > 0$ (no matter how small r is). In other words, a sequence converges to x if it gets close to x and stays close as n gets larger and larger. Sometimes it happens that a sequence gets close to a point every once in a while but doesn't stay close—it hops around anywhere, but every now and then it comes back near the given point. This kind of behavior is also important.

5.7. Definition. Let $\{x_n : n \in \mathbf{Z}^+\}$ be a sequence in (X, d). A point $x \in X$ is an **accumulation point** of the sequence if given any positive real number r (no matter how small) and given any integer N (no matter how large), then there exists a positive integer $n \geq N$ such that $x_n \in S_r(x)$.

Thus x is an accumulation point of a sequence $\{x_n : n \in \mathbf{Z}^+\}$ in (X, d) if for any $r > 0$ (no matter how small), there exist arbitrarily large integers n such that $x_n \in S_r(x)$. Instead of the phrase "given any integer N (no matter how large) there exists an $n \geq N$ such that $x_n \in S_r(x)$," we say for convenience that "x_n is *frequently* in $S_r(x)$." Thus the definition of accumulation point could be written as:

A point $x \in X$ is an **accumulation point** of a sequence $\{x_n : n \in \mathbf{Z}^+\} \subset (X, d)$ if given $r > 0$ (no matter how small), x_n is frequently in $S_r(x)$.

5.8. Exercises.
1) If a sequence converges to a point x, then x is an accumulation point of the sequence. (So "ultimately" implies "frequently.")

2) If x is an accumulation point of a sequence, the sequence need *not* converge to x. (So "frequently" does not imply "ultimately.")

3) A subset F of a metric space X is closed if and only if whenever $x \in X$ is an accumulation point of a sequence of points of F, then $x \in S$. [*Hint*: See Theorem 5.5.]
4) State precisely what it means when the point $x \in X$ is *not* an accumulation point of the sequence $\{x_n : n \in \mathbf{Z}^+\} \subset X$.

6. CONTINUITY

Like convergence, continuity of functions is subtle but extremely important. In fact, continuity is probably the single most important concept in all mathematics. Recall that a function is a way to get from one set to another, or, speaking topologically, a function is a way to transform one topological space into another. When the function is continuous, most of the important properties that the domain space has (like being all in one piece, for example) are preserved under the transformation, so that the image space also has these properties. This preservation of important properties is of the utmost importance in topology, as we will see very often as we go on.

You probably remember from calculus that "a function (from $D \subset \mathbf{R}$ to $\mathbf{R}$) is continuous if you can draw its graph without lifting your pencil from the paper; the graph of a continuous function has no holes or breaks in it." This is easy to understand but it is hardly very precise (and is not even true in all cases as we will see). It *is* true if the domain of the function is an interval, and in most calculus courses, it is made precise by an "ϵ-δ" definition, as follows.

6.1. Definition. Let $D \subset \mathbf{R}$ and let $f: D \to \mathbf{R}$. Then f is **continuous at a point** $x_0 \in D$ if for any positive real number ϵ there exists a positive real number δ such that $|f(x) - f(x_0)| < \epsilon$ whenever $x \in D$ and $|x - x_0| < \delta$. The function f is then said to be **continuous** (on D) if it is continuous at every point of D.

This definition is *very* subtle, as most first-year calculus students know. It makes the statement mentioned above about the graph of the function mathematically rigorous. A better way to think of it though is in terms of "closeness": the definition says that a continuous function preserves closeness, in the sense that points that are close to x_0 in the domain of a continuous function are sent to points that are close to $f(x_0)$ in the range. The idea of closeness is made precise by asking and answering a question, namely: how close to x_0 is close enough for points x to be so that their images $f(x)$ are within a prescribed distance (within ϵ) of $f(x_0)$? The fact that the question can be answered (i.e., that a δ can be shown to exist), no matter how small the prescribed distance ϵ is, means that the function is continuous at the point x_0—it preserves closeness.

For example, consider the function from **R** to **R** given by

$$f(x) = \begin{cases} 1 & \text{if } x \geq 2 \\ -1 & \text{if } x < 2. \end{cases}$$

Its graph looks like

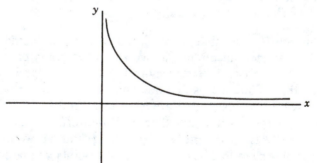

This function is clearly not continuous at $x = 2$ (it *is* continuous every-where else). Saying that it is not continuous at $x = 2$ says that there are points that are close to $x = 2$ that are not sent to points that are close to $f(2) = 1$. More precisely, it must be possible to give an $\epsilon > 0$ for which there is no corresponding δ—for some $\epsilon > 0$ it must be true that no matter how small an open interval we pick around $x = 2$, there will be some points in this interval whose images are more than ϵ away from $y = 1$ (with y-distances measured on the y-axis). Finish up the details by doing Problem 2 in Exercises 6.2 below.

The function defined above is not continuous at $x = 2$ because it fails to preserve "closeness" at $x = 2$. To pursue the "closeness" idea a little further, consider the function defined by

$$g(x) = 1/x \quad \text{if } x > 0.$$

This function *is* continuous on its domain (0 is not in the domain!) Its graph looks like

Since it is continuous, this function must preserve closeness. But "close" is a relative term. For example, if $x = \frac{1}{10}$, then $g(x) = 10$; the points in the interval $(\frac{1}{11}, \frac{1}{9})$ are pretty close to $x = \frac{1}{10}$, but g spreads these points over the interval (9, 11) on the y-axis, and there are points in (9, 11) almost a whole unit away from $g(\frac{1}{10}) = 10$. As another example, the points $x = 0.004$ and $x = 0.005$ might be said to be *very* close together (the distance between them is only 0.001 unit); but $g(0.004) = 250$ and $g(0.005) = 200$, so the images under g of these two nearly indistinguishable points are 50 units apart! And things can look even worse than this: the points $x = 10^{-6}$ and $x = 10^{-7}$ are so close together that you would need a pretty powerful microscope to tell them apart. But their images under this continuous function are *9,000,000* units apart!

But this function g *is* continuous. Prove it by doing Problem 3 of Exercises 6.2, and think about it enough to convince yourself that g does preserve closeness: for a given $y_0 = g(x_0)$, it is true that $g(x)$ will be as close as you want to $g(x_0)$ provided that x is close enough to x_0. But for this function, close enough to x_0 can mean *very, very* close.

6.2. Exercises.

1) State precisely what it means when a function f is *not* continuous at a point x_0 in its domain.
2) Prove that the function defined by

$$f(x) = \begin{cases} 1 & \text{if } x \geq 2 \\ -1 & \text{if } x < 2 \end{cases}$$

is not continuous at $x = 2$, but is continuous everywhere else.
3) Prove that the function g defined by $g(x) = 1/x$ for $x \in \mathbf{R}$ and $x > 0$ is continuous on its domain.

We want to get away from this elementary calculus point of view and look at continuity in more generality. For convenience, we repeat Definition 6.1 here.

A function $f: D \subset \mathbf{R} \to \mathbf{R}$ is **continuous at** $x_0 \in D$ if for any positive real number ϵ there exists a positive real number δ such that $|f(x) - f(x_0)| < \epsilon$ whenever $|x - x_0| < \delta$. The function is **continuous** (on D) if it is continuous at every point of D.

This definition talks about closeness in terms of the metric on the real line (recall that the distance between two real numbers x and y is given by the metric d, where $d(x, y) = |x - y|$). A generalization to arbitrary metric spaces should be obvious. (Recall the definition of a metric space given in Section 4; if d is a metric on X, then the *distance* between two points x_1 and x_2 of X is given by $d(x_1, x_2)$.)

6.3. Definition. Let (X, d_1) and (Y, d_2) be metric spaces, and let $f: D \subset X \to Y$. Then f is **continuous at** $x_0 \in D$ if for any positive real number ϵ there exists a positive real number δ such that $d_2(f(x), f(x_0)) < \epsilon$ whenever

$d_1(x, x_0) < \delta$. The function f is **continuous** (on D) if it is continuous at every point of D.

Notice that this definition is the same as Definition 6.1 when (X, d_1) and (Y, d_2) are both the real line with its usual metric topology, so it does generalize Definition 6.1. (We say that a statement A *generalizes* a statement B if they say the same thing when they both apply, but A applies to more situations than B does.) Notice also that in Definition 6.3, the distance between x and x_0 in X is measured using the metric d_1, whereas the distance between $f(x)$ and $f(x_0)$ in Y is measured using the metric d_2.

Recall that if r is a positive real number, then an *r-ball* about a point p_0 in a metric space is the set of all points in the space whose distance from p_0 is less than r. Definition 6.3 can then be restated as:

Let (X, d_1) and (Y, d_2) be two metric spaces and let $f: D \subset X \to Y$. Then f is **continuous at** $x_0 \in D$ if given any positive real number ϵ there exists a positive real number δ such that $f(x) \in S_\epsilon(f(x_0))$ whenever $x \in S_\delta(x_0)$, so that $f(S_\delta(x_0)) \subset S_\epsilon f(x_0))$. The function f is then said to be **continuous** (on D) if it is continuous at each point of D.

This definition says precisely what continuity is for a function between two metric spaces. But we want to keep the intuitive idea in mind: a function f is continuous if it preserves closeness—it doesn't tear the domain apart by sending two points that are close together to points that are far apart (all of these terms being relative of course).

The closeness idea might remind you of sequences (look at Theorem 5.5 again). In fact, in metric spaces, continuity can be characterized in terms of sequences as the following theorem shows. This theorem is very important.

6.4. Theorem. Let (X, d_1) and (Y, d_2) be metric spaces and let $f: D \subset X \to Y$. Then f is continuous at a point $x_0 \in D$ if and only if whenever $\{x_n : n \in \mathbf{Z}^+\}$ is a sequence in D that converges to x_0, then the sequence $\{f(x_n) : n \in \mathbf{Z}^+\}$ converges to $f(x_0)$ in Y.

[*Hint for proof*: Think! Read the definitions of convergence and continuity very carefully.]

Thus a function from one metric space to another is continuous if and only if it preserves convergent sequences.

6.5. Exercises.
1) Let S be a sequence in $\mathbf{R}$, and let $\mathbf{Z}^+$ have the discrete topology that it gets by saying that every point in $\mathbf{Z}^+$ is open. Prove that S is continuous.
2) Let p be a point, and let $Y = \{p\}$. Make Y into a topological space by declaring that the sets $\emptyset$ and $Y = \{p\}$ are open (don't try to prove that Y is a topological space—we don't have the necessary defini-

tion yet; we will prove that it is a topological space in the next chapter). Let $f: \mathbf{R} \to Y$ be defined by $f(x) = p$ for all $x \in \mathbf{R}$. Prove that f is continuous. This is a special case of a theorem that says that any constant function is continuous.

3) Let $f: \mathbf{R} \to \mathbf{R}$ be defined by

$$f(x) = \begin{cases} 1, & \text{if } x \text{ is rational} \\ -1, & \text{if } x \text{ is irrational.} \end{cases}$$

Prove that f is not continuous on $\mathbf{R}$. Is f continuous at any point of $\mathbf{R}$?

4) Let $f: \mathbf{R}^+ \to \mathbf{R}$ be defined by

$$f(x) = \begin{cases} 1/q, & \text{if } x = p/q \text{ in lowest terms} \\ 0, & \text{if } x \text{ is irrational} \end{cases}$$

Is f continuous? Is f continuous at any point of $\mathbf{R}^+$?

7. HOMEOMORPHISM

We have said that continuity is important in topology because certain properties that a topological space might have are preserved when that space is transformed into another by a continuous function. For example, a circle can be transformed into a square (as in the Introduction) in a continuous way, and everyone will agree that these two topological spaces look very much alike. As another example, we will show later that it is *impossible* to have a continuous function which sends an interval onto two disjoint intervals: any function that transforms a single interval into two disjoint intervals would have to rip its domain interval apart, and ripping is not a continuous operation (unless the "rip" is "repaired" before the transformation is completed. For example, there is a continuous function from the knot (*A*) below onto the circle (*B*), even though the knot must be torn, untwisted, and then repaired to transform it onto the circle. The important thing for such a function to be continuous is that the cut made in the knot is repaired before the function is finished).

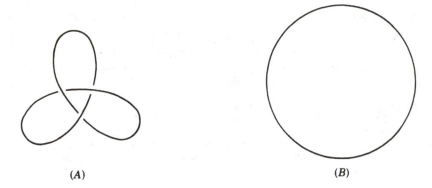

(A)   (B)

Unfortunately, though, continuity of a function is not enough to insure that *all* important properties of the domain space are preserved when it is transformed onto its range. Look at Problem 2 in Exercises 6.5. It shows that a function that transforms the whole real line into a single point is continuous. (Of course it is. You can just squeeze the line together to a point, and you don't have to rip it apart to do this.) But it is obvious that the line and a single point are very different looking spaces. The problem is that even though we can transform the line continuously into a point, we cannot undo this transformation and have a function from the point back onto the line (Why not?). In the case of a circle and a square, the continuous changes that are made to go from one space to the other can be reversed, and the range space can be continuously transformed back to what it came from. (The same is true with changing the doughnut into the coffee cup—the transformation can be reversed.)

A continuous function, then, preserves *some* properties that a space has, but it does not necessarily preserve them all. In order that *all* important properties of two spaces be the same, a function between them needs to be more than just continuous. Consider the following definition.

7.1. Definition. Let X and Y be topological spaces. A function $f: X \rightarrow Y$ is called a **homeomorphism** or a **topological transformation** if it is 1-1, continuous, and $f^{-1}: f(X) \rightarrow X$ is also continuous.

When $f: X \rightarrow Y$ is a homeomorphism, X and its image, $f(X)$ are said to be **homeomorphic,** and we write $X \cong f(X)$. In case f is also *onto* Y (so that $f(X) = Y$), then $X \cong Y$.

We have said that the important properties that a topological space has are shared by any other topological space that is homeomorphic to it. We call such properties *topological*. Formally,

7.2. Definition. A property P of a topological space X is a **topological property** if whenever Y is a topological space such that $Y \cong X$, then Y also has property P.

There are many topological properties, as we will see. For subsets of the real line, two of them are the property of being all in one piece (being *connected*), and the property of being closed and bounded (being *compact*).* We will investigate connectedness and compactness in some detail and in more generality in later chapters. (We should mention that these two prop-

---

* Actually, connectedness and compactness are both preserved by continuous functions. In other words, if $f: X \rightarrow Y$ is continuous and onto, then if X is connected, so is Y, and if X is compact, Y is also compact. However, it is possible for $f: X \rightarrow Y$ to be continuous and onto and for Y to be connected (compact) while X is not connected (compact). If we want *both* X and Y to be connected (compact), then there must be a homeomorphism between X and Y—continuity alone is not enough.

erties are topological whether the real line is involved or not, but their descriptions may have to be changed from those given above when we are dealing with spaces other than **R**.) Some properties that are *not* topological include size (the length of an interval, for example: we can stretch or shrink an interval without changing it topologically, as long as we don't shrink it too much), "degree of curveness" (a circle, an ellipse, a square, and a triangle, for example, are all the same topologically), and order (the discrete space $\mathbf{Z}^+$ is homeomorphic to the discrete space of negative integers, even though their orderings are exactly opposite).

We establish a few examples a little more precisely in the following exercises.

7.3. Exercises.

1) Consider a closed interval $[a, b]$ in the real line with a and b both finite, and $a < b$. Make this interval into a topological space by giving it the metric topology induced by the absolute value metric $d(x, y) = |x - y|$, as in Section 3. (Recall that intervals of the form $[a, x)$ and $(y, b]$, where x and y are between a and b, are open in this space.)

   Prove that $[0, 1] \cong [2, 4]$. This shows that length is *not* a topological property—it is not necessarily preserved by a homeomorphism. [*Hint:* Put $[0, 1]$ on the x-axis and $[2, 4]$ on the y-axis in the plane; think about a linear function.]

2) Consider the open unit interval $(0, 1)$ as a topological space by giving it the metric topology induced by the absolute value metric $d(x, y) = |x - y|$ as in Section 3. Prove that $(0, 1) \cong \mathbf{R}$. This shows again that length is not a topological property—it is not necessarily preserved by a homeomorphism. In fact, we have here an interval of finite length homeomorphic to one whose length is infinite! [*Hint:* An appropriate modification of the function $f(x) = \tan x$ or something whose graph looks like the graph of the tangent function might be a good thing to look at.]

3) We can think of a circle or a square as the closed unit interval with its end points identified into a single point. The topology on the circle (or the square) is just what you would expect: a subset of the circle (or the square) is open if and only if it contains an open interval around each of its points (except that intervals are curved on the circle and may be bent around the corners on the square).

   Prove that the unit circle centered at the origin is homeomorphic to a square with unit perimeter centered at the origin. [*Hint:* Think of the origin as a light bulb: project from the origin.] Can you generalize this to arbitrary circles and squares in the plane? How about a triangle?

4) It was stated in the introduction that a circle is not topologically the same as a straight line segment. The reason for this is not because the circle is curved and the line is not ("curveness" is *not* topological). Rather, it is because of the points of the line that are not end points.

   a) Let $[a, b]$ be a closed interval on the real line with a and b both finite and $a < b$ (so $[a, b]$ is a line segment). Convince yourself that the property of not being an end point is a topological property; in other words, if $h:[a, b] \rightarrow Y$ is a homeomorphism, then for $x \neq a$ and $x \neq b$, $h(x)$ cannot be an "end point" of Y, in the sense that if $h(x)$ is removed from Y, then the resulting space $(Y - \{h(x)\})$ consists of at least two pieces. The technical name for "non-end point" is **cut point**, because the removal of a cut point from a space cuts it into at least two pieces. Hence, in homeomorphic spaces, cut points must correspond to cut points.

   b) Convince yourself that a line segment and a circle are not homeomorphic. [*Hint:* What happens when you remove a point from a circle? How many cut points does a circle have?]

5) We said in the discussion of Cartesian product (Section 3 of Chapter I) that X is not a subset of $X \times Y$. However, X is a "topological subset" of $X \times Y$ in the sense that X is homeomorphic to a subset of $X \times Y$. Illustrate this in the case of $\mathbf{R} \times \mathbf{R}$ by convincing yourself that $\mathbf{R}$ is homeomorphic to the x-axis $(= \mathbf{R} \times \{0\})$ when the x-axis is given a topology as follows: a subset U of the x-axis is open if and only if for each point $(x, 0) \in U$, there is an $r > 0$ such that $S_r(x, 0) \cap (x\text{-axis})$ is contained in U. Is $\mathbf{R}$ homeomorphic to any other subset of $R \times R$ when that subset is topologized analogously?

Deciding whether or not two topological spaces are homeomorphic is very important in topology because it is often the case that we know some important properties that one space has while we know very little about the other. If it turns out that the differences between the two are merely superficial—if the two are in fact homeomorphic—then the important things that we know about the first space, we also know about the second. For example, some people get very nervous about infinity (∞): since the real line is homeomorphic to the open unit interval, these people can avoid ∞ altogether when they think about the line topologically. Of course there are many more important examples, as we will see as we go along. One such example that we can consider now concerns the following subsets of the plane, each of which is given a topology as follows: if X denotes the subset in question, then a subset U of X is open if for any point $x \in U$, there is an $r > 0$ such that $S_r(x) \cap U \subset U$. Actually, to visualize a homeomorphism between the two spaces, you need not be concerned too much with the topology that they have. Of course, if you were asked to *prove* that they are homeomorphic, then the topologies would be of the utmost importance. The two subsets look like:

(A)

(B)

Space (A) is nothing but the graph of the function $y = \sin (1/x)$ together with its "limit line," the line from -1 to 1 on the y-axis. Since it is mostly nothing but the graph of a continuous function, we should know a lot about space (A). But what is space (B)? It spirals around a line and might look pretty strange, but it turns out that spaces (A) and (B) are in fact homeomorphic. Try to visualize how you could transform one into the other. (You might have to lift one of them out of the plane during the transformation.)

A homeomorphism is a continuous function between two topological spaces X and Y with some special properties. Another important kind of continuous function is a function from a topological space to a subset of that space, which leaves the subset alone. It is defined for **R** as follows.

7.4. Definition. Let $A \subset \mathbf{R}$ and let both A and **R** have the metric topology induced by the usual absolute value metric. Then A is a **retract** of **R** if there is a continuous function $r: \mathbf{R} \to A$ such that $r(a) = a$ for all $a \in A$. The function r is called a **retraction** of **R** onto A.

7.5. Exercises.
1) Show that for any point $p \in \mathbf{R}$, $\{p\}$ is a retract of **R**.
2) Show that $(0, 1)$ is a retract of **R**.
3) Show that $A \subset \mathbf{R}$ is a retract of **R** if and only if the identity function $\mathrm{id}_A : A \to A$, defined by $\mathrm{id}_A(a) = a$ for $a \in A$, can be *extended* over **R** in the following sense. A function $f: A \to A$ can be *extended* over **R** if there is a continuous function $F: \mathbf{R} \to \mathbf{R}$ such that $F(a) = f(a)$ for all $a \in A$.
4) Do you think that $\{0, 1\}$ is a retract of **R**? Why or why not?

8. TOPOLOGICAL MANIFOLDS

When people used to think that the world was flat, they had good reason because *locally* the world *is* flat (except for hills and valleys of course—essentially flat anyway). By *locally flat* we mean that at any given point, it is possible to have a region around that point which is topologically the same as some region in the plane, and on the surface of the earth, this is indeed possible. The surface of the earth is a *topological 2 manifold:* every point on it has an open set around it which is homeomorphic to an open set in E^2, Euclidean 2-space—the plane. (The mistake of those who thought the earth was flat of course was in assuming that the fact that the earth is *locally* like the plane meant that it *was* the plane.)

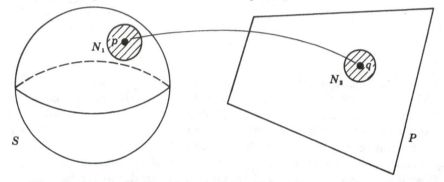

The shaded open sets, N_1 containing the point p on the sphere (S) and N_2 containing the point q in the plane (P) are homeomorphic, because if you "flatten out" N_1 you get N_2. This is true for every point on the sphere, so it looks locally exactly like the plane—it is a topological 2-manifold.

Clearly, then, a topological 2-manifold is a space that is essentially 2-dimensional. But according to our description so far, the disk $\{(x, y):x^2 + y^2 \leq 1\}$ would not be a topological 2-manifold even though it is not only *essentially* 2-dimensional but in fact *is* 2-dimensional. The reason why this disk fails our criterion to be a topological 2-manifold is because of the points on its edge—its *boundary points*. Convince yourself that none of these boundary points is contained in an open subset of the disk which is homeomorphic to an open subset in the plane, because a boundary point is always on the "edge" of any subset of the disk that contains it. Recall, though, that when we considered a closed interval $[a, b] \subset \mathbf{R}$ as a topological space by using the absolute value metric, we showed that sets of the form $[a, x)$ and $(y, b]$ are open subsets of the interval, even though they are not open subsets of the line. The same thing happens with the disk. The disk is a metric space with topology induced by the usual distance formula in the plane, and using this metric, if (x, y) is a boundary point of the disk (i.e., if

$x^2 + y^2 = 1$), then any set of the form $S_r(x) \cap D$ (where D is the disk) is open in the disk. But a set of this form is not homeomorphic to any open subset of the plane. It is, however, homeomorphic to an open subset of the upper half plane, $\{(x, y): x, y \in \mathbf{R} \text{ and } x \geq 0\}$, when this set is given the topology induced by the usual distance formula in the plane.

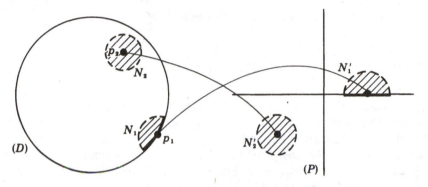

(D) is the disk $\{(x, y): x^2 + y^2 \leq 1\}$ and (P) is the plane. The interior point p_2 of (D) is contained in the open set N_2 in (D) which is homeomorphic to the open set N_2' in (P). The boundary point p_1 in (D) is contained in the open set N_1 of (D) which is homeomorphic to the open set N_1' in the upper half plane $\{(x, y): x, y \in \mathbf{R} \text{ and } x \geq 0\}$.

Thus we should modify our description of a topological 2-manifold so that spaces which are essentially 2-dimensional will be topological 2-manifolds even if they have boundary points. The key to this modification is in the discussion given above. Since every boundary point of the disk is contained in an open subset of the disk which is homeomorphic to an open subset in the upper half plane, as in the picture, we say that a space X is a topological 2-manifold if every point of X is contained in an open set which is homeomorphic to an open subset of the plane, or to an open subset of the upper half plane.

Analogous to topological 2-manifolds are topological 1-manifolds. As might be expected, a topological 1-manifold is a space that is essentially 1-dimensional: any point in a topological 1-manifold is contained in an open set that is homeomorphic to an open subset of $\mathbf{R}^1$ $(=\mathbf{R})$ or to an open subset of $[0, \infty)$. (In the latter case, if the point is not contained in an open set homeomorphic to an open subset of $\mathbf{R}$, then it is called a *boundary point*; otherwise, it is an *interior point*.) Thus a line segment $[a, b]$, for example, is a topological 1-manifold *with boundary*, while a circle is a topological 1-manifold *without boundary*.

Before we can discuss manifolds of dimension > 2, we need higher dimensional Euclidean spaces.

8.1. Definition. Let n be a positive integer. **Euclidean n-space,** E^n, is the set

$$\mathbf{R}^n = \underbrace{\mathbf{R} \times \mathbf{R} \times \cdots \times \mathbf{R}}_{n} = \left\{ (x_1, x_2, \cdots, x_n) : x_i \in \mathbf{R} \right\}$$

with the metric topology induced by the metric*

$$d(x, y) = \sqrt{\sum_{i=1}^{m} (x_i - y_i)^2},$$

where $x = (x_1, x_2, \cdots, x_n)$ and $y = (y_1, y_2, \cdots, y_n)$.

(Note that $\mathbf{R}^n$ is the *set* of n-tuples of real numbers, and E^n is the metric *space* obtained from the set $\mathbf{R}^n$ by giving it the topology induced by the generalized distance formula, although we deviate from this convention in the case of dimension 1, and usually just use $\mathbf{R}$ to denote the space E^1.) Thus $E^1 = \mathbf{R}$ is the line, E^2 is the plane and E^3 is ordinary 3-dimensional space, as we have already seen. For $n > 3$, E^n is defined in Definition 8.1, but what it looks like can only be imagined.

8.2. Definition.
  a) A **topological n-manifold** M **without boundary** is basically† a space in which every point is contained in an open set that is homeomorphic to an open subset of E^n.
  b) A **topological n-manifold** M **with boundary** is basically a space in which every point satisfies (a), but there exist points of M that do not satisfy (a), and each of these points (called **boundary points**) is contained in an open set homeomorphic to an open subset of $\{(x_1, x_2, \cdots, x_n) : x_i \in \mathbf{R} \text{ and } x_1 \geq 0\}$.

8.3. Exercises.
  1) Determine all topologically different connected 1-manifolds that you can (*connected* means all in one piece). Try to determine any essential differences between them.
  2) Is a space that looks like a Y a topological 1-manifold? Why or why not?
  3) Give as many different examples of connected topological 2-manifolds as you can and indicate why you think they are different.
  4) Find out what a Möbius strip and a Klein bottle are. Do you want to add anything to your answers to Problems 1 or 3 above?

---

* Proving the triangle inequality for this metric is quite messy. Accept the fact that it is a metric for now. We will prove it later.

† We have to say "basically" here because a manifold must have some properties that we have not discussed. All of the spaces that you will think of as manifolds will have these properties, so it is really no problem for now.

chapter four

Topological Spaces and Basic Topological Concepts in General

In this chapter we will define topological spaces in general and will look at some topological spaces and some topological ideas which are probably not as familiar as those that we have seen up to now. To help keep things straight, you should keep the familiar spaces (especially the real line) in mind when you study the definitions and prove the theorems, and you should draw pictures of the situation at hand. But remember that a picture is not a proof.

1. DEFINITION OF A TOPOLOGICAL SPACE

In Theorem 1.3 of Chapter 3 we proved that both **R** and $\emptyset$ are open in **R**, that the union of any collection of open subsets of **R** is open, and that the intersection of any finite collection of open subsets of **R** is open. Since a similar theorem is satisfied by many other familiar spaces, as we saw in the exercises, we abstract these properties and make the following definition.

1.1. Definition. Let X be a set. A **topology** on X is a collection of subsets $\mathfrak{I}$ of X that satisfies the following three conditions:
1) $\emptyset$ and X belong to $\mathfrak{I}$.
2) If $\mathfrak{A} \subset \mathfrak{I}$, then $\cup\mathfrak{A} \in \mathfrak{I}$.
3) If $\mathfrak{B}$ is finite and $\mathfrak{B} \subset \mathfrak{I}$, then $\cap\mathfrak{B} \in \mathfrak{I}$.

The pair $(X, \mathfrak{I})$ is called a **topological space,** and the subsets of X that belong to $\mathfrak{I}$ are called **open sets in** $(X, \mathfrak{I})$, or, sometimes, if it doesn't lead to confusion, just **open subsets of** X, or even just **open sets.**

Definition 1.1 can be stated in words as:

In a topological space,
1) The empty set and the whole space are open sets,

2) The union of any collection of open sets is an open set, and

3) The intersection of any finite collection of open sets is an open set.

When we are dealing with a topological space, the particular topology that the space has is often not important, and results that we obtain in this case are valid for any topological space. In such a situation, we will often omit the $\mathfrak{I}$, and simply say that X is a topological space. When the particular topology on X does matter, we will specify it by saying that our space is $(X, \mathfrak{I})$. (Of course, other letters can be used, both for the set X and the topology $\mathfrak{I}$ that X has.)

As you might expect,

1.2. Definition. Let X be a topological space. A subset $F \subset X$ is closed in X if its complement $(X - F)$ is open in X.

If the topology on X is $\mathfrak{I}$, then we could say that $F \subset X$ is closed in X if and only if $X - F \in \mathfrak{I}$. This is the same as saying that $X - F$ is open in X.

2. EXAMPLES

We have seen many examples of topological spaces. Keep them in mind while we look at some that are a little more abstract.

2.1. The Indiscrete Topology. This is the simplest topology that it is possible to assign to a set because the only open sets are the whole space itself and the empty set. Formally, for any set X, the **indiscrete topology** on X is

$$\mathscr{I} = \{X, \emptyset\}.$$

Even if X is the real line, we can give it the indiscrete topology instead of the usual one that we defined in Chapter 3. Thus *a set can have more than one topology assigned to it*. When **R** has the indiscrete topology, nothing is open except **R** and $\emptyset$—in particular, *ordinary open intervals are not open in* **R** *when it has the indiscrete topology*.

2.2. The Discrete Topology. The smallest (in the sense of having the fewest number of open sets) topology that we can give to a set is the *indiscrete* topology. In the *discrete* topology, *every* subset of X is open, as we saw in Chapter 3, so the discrete topology is the largest topology that we can give to a set. Formally, if X is any set, the **discrete topology** on X is defined by

$$\mathfrak{D} = \mathcal{P}(X) = \{S : S \subset X\}.$$

When we give the discrete topology to **R**, we get a topological space that is different from both $(\mathbf{R}, \mathscr{I})$ and $(\mathbf{R}, \mathfrak{U})$, where $\mathfrak{U}$ is the usual topology induced by the absolute value metric as defined in Chapter 3. In particular, ordinary closed intervals are *open* in $(\mathbf{R}, \mathfrak{D})$.

2.3. Topologies on a 2 Point Set. Let $X = \{0, 1\}$, a set consisting of two points. There are three distinct topologies that can be assigned to X, two of which we have already discussed.

a) The indiscrete topology on $\{0, 1\}$ is

$$\mathfrak{g} = \{\emptyset, \{0, 1\}\}.$$

b) The discrete topology on $\{0, 1\}$ is

$$\mathfrak{D} = \{\emptyset, \{0\}, \{1\}, \{0, 1\}\}.$$

c) A topology between the indiscrete and the discrete topologies that can be assigned to $\{0, 1\}$ is

$$\mathfrak{s} = \{\emptyset, \{0\}, \{0, 1\}\}.$$

The resulting topological space $(\{0, 1\}, \mathfrak{s})$ is known as **Sierpinski space.***

2.4. The Finite-Complement Topology. If X is a set, we can assign a topology to X by declaring that a subset U of X is open if and only if its *complement* (relative to X) is finite, or if U itself is empty. The topology that we get is called the **finite-complement topology** (sometimes called the **co-finite** topology) and is the collection

$$\{U : U \subset X \quad \text{and} \quad X - U \quad \text{is finite or} \quad U = \emptyset\}.$$

When the finite-complement topology is assigned to the real line, the resulting topological space is different from all of the previous spaces that we have built on the real line. Thus a set can have *many* different topologies defined on it.

2.5. The Sorgenfrey Line.† This is yet another topology that can be assigned to the real line. It is defined by saying that a subset U of the real line is open in the **Sorgenfrey topology** on the line if and only if for any point $x \in U$, there exist real numbers a and b such that $x \in [a, b) \subset U$. (Note that the empty set is open *vacuously* in the Sorgenfrey topology: since there are no points in the empty set, the defining condition is automatically satisfied for the empty set.) The Sorgenfrey topology is different from all of the other topologies that we have defined on the real line; in particular, "half-open" intervals of the form $[a, b)$ are *open* in this topology!

---

* After Waclaw Sierpinski (1882–1969), one of the many brilliant twentieth Century Polish mathematicians.
† After Robert Sorgenfrey (1915–   ).

2.6. Exercises.

1) Let X be a set. Verify that the indiscrete topology, the discrete topology and the finite-complement topology are in fact topologies on X.

2) a) Verify that Sierpinski space is a topological space.

   b) We said that there are only three different topologies that can be assigned to the 2 point set $\{0, 1\}$. Is the collection of sets $\{\emptyset, \{1\}, \{0, 1\}\}$ one of these three topologies on $\{0, 1\}$?

   c) What is $\{0, 1\}$ with the finite-complement topology?

3) List all topologies that can be assigned to a 3 point set.

4) Verify that the Sorgenfrey topology defined on the real line is in fact a topology. Is the interval $(0, 1)$ open in this topology? How about $(0, 1]$? Is $[0, 1]$ closed?

5) Consider the topological spaces $(\mathbf{R}, \mathcal{I})$, $(\mathbf{R}, \mathcal{D})$, $(\mathbf{R}, \mathcal{U})$, $(\mathbf{R}, \mathcal{S})$ and $\mathbf{R}$ with the finite-complement topology (where $\mathcal{U}$ denotes the usual topology on $\mathbf{R}$ as in Chapter 3).

   a) If $p \in \mathbf{R}$, is $\{p\}$ open in any of these spaces? Which ones?

   b) If $p \in \mathbf{R}$, is $\{p\}$ closed in any of these spaces? Which ones?

   c) In which of these spaces is (a, b) open? $[a, b)$? $(a, b]$? $[a, b]$?

   d) Is the set $\{x \in \mathbf{R} : x \neq 1/n\}$ open in any of these spaces? Is it closed in any of them?

   e) Is the set $\{x \in \mathbf{R} : x \neq 1/n$ and $x \neq 0\}$ open in any of these spaces? Is it closed in any of them?

6) Consider the spaces of Problem 5 above again, together with the three spaces that can be defined on $\{0, 1\}$.

   a) In which of these spaces is the following statement true: If x and y are two distinct points in the space then either there exists an open set U such that $x \in U$ and $y \notin U$, or there exists an open set V such that $y \in V$ and $x \notin V$. (A space for which this statement holds is called a T_0-**space**.)

   b) In which of these spaces is the following statement true: If x and y are two distinct points in the space, then there exists an open set U such that $x \in U$ and $y \notin U$, and there exists an open set V such that $y \in V$ and $x \notin V$. (A space for which this statement holds is called a T_1-**space**.)

   c) In which of these spaces is the following statement true: If x and y are two distinct points in the space, then there exist open sets U and V such that $x \in U$, $y \in V$ and $U \cap V = \emptyset$. (A space for which this statement holds is called a T_2-**space** or a **Hausdorff** space.*)

* After Felix Hausdorff (1868–1942), one of the founders of the study of topology as such.

7) The three conditions defined in Problem 6 above are called **separation axioms** because they tell how points can be separated from one another in various topological spaces. Show that every T_2-space is a T_1-space, and that every T_1-space is a T_0-space, and give an example of a T_0-space that is not a T_1-space, and an example of a T_1-space that is not a T_2-space. We say, then, that "T_2 is stronger than T_1, and T_1 is stronger than T_0." In a T_2-space, any two points can be separated by disjoint open sets, in a T_1-space, any two points can be separated by open sets, but the separating open sets may overlap, and in a T_0-space, given any two points, at least one of them can be separated from the other by an open set. It turns out that topological spaces are not of much practical use unless they are at least T_2, and, in this book, all of the spaces that we will concentrate on will be T_2-spaces.

8) A topological space X is said to be **metrizable** if a metric can be defined on X so that a set is open in the metric topology induced by this metric if and only if it is open in the topology that is already on the space.

   a) Let X be a set with more than one point. Prove that $(X, \mathcal{I})$ is *not* metrizable. Thus the indiscrete topology on a set with more than one point is an example of a topological space that is *not* a metric space.

   b) Let X be a set. Define a function from $X \times X = \{(x, y) : x, y \in X\}$ to **R** by

$$d(x, y) = \begin{cases} 1 & \text{if } x \neq y \\ 0 & \text{if } x = y. \end{cases}$$

Prove that d is a metric on X. What is the metric topology induced by d?

3. CONTINUITY

For convenience, we repeat the definition of continuity previously given for metric spaces (Definition 6.3 in Chapter 3), as it was finally phrased.

3.1. Definition. Let (X, d_1) and (Y, d_2) be two metric spaces and let $f : D \subset X \to Y$. Then f is **continuous at** $x_0 \in D$ if given any positive real number ϵ there exists a positive real number δ such that $f(x) \in S_\epsilon(f(x_0))$ whenever $x \in S_\delta(x_0)$. In other words, f is continuous at $x_0 \in D$ if given $\epsilon > 0$ there exists $\delta > 0$ such that $f(S_\delta(x_0)) \subset S_\epsilon(f(x_0))$. The function f is then said to be **continuous** (on D) if it is continuous at each point of D.

As we noted in the discussion in Chapter 3, this definition makes precise the intuitive idea of a function being continuous if it preserves "closeness":

when $f: X \rightarrow Y$ is continuous, points close to x_0 in X are sent by f to points that are close to $f(x_0)$ in Y (how close being relative of course).

It might seem that metric spaces are necessary in any discussion of continuity because we need a metric (a distance function) to measure the distance between points to see how close together they are. But it turns out that it is not the metrics themselves that are important to continuity; rather, the important things are the topologies induced by these metrics, and the fact that these topologies come from metrics is really of no consequence once we have them. In fact, we can discuss continuity of a function between two metric spaces without mentioning the metrics at all, as the following theorem shows. (We assume for now that the domain of f is all of X (rather than a subset $D \subset X$) in order to avoid irrelevant complications.)

3.2. Theorem. Let (X, d_1) and (Y, d_2) be metric spaces and let $f: X \rightarrow Y$. Then f is continuous at $x_0 \in X$ if and only if whenever V is an open subset of Y with $f(x_0) \in V$, then there exists an open subset U of X such that $x_0 \in U$ and $f(U) \subset V$.

[*Hint for proof*: Remember that a set is open in a metric space if and only if it contains an open r-ball around each of its points.]

Theorem 3.2 suggests the following important generalization of the idea of continuity to arbitrary topological spaces.

3.3. Definition. Let X and Y be topological spaces and let $f: X \rightarrow Y$. Then f is **continuous at** $x_0 \in X$ if whenever V is an open subset of Y with $f(x_0) \in V$ then there exists an open subset U of X with $x_0 \in U$ and $f(U) \subset V$. The function f is then said to be **continuous** (on X) if it is continuous at every point of X.

Theorem 3.2 shows that this new definition is indeed a generalization of the ϵ-δ definition, because when the two spaces involved are metric spaces, the new definition is equivalent to the ϵ-δ definition. (Recall that a statement A is a generalization of a statement B if the two statements say the same thing whenever statement B makes sense, but A applies to more cases than B does. In our situation here, the statements of Definition 3.3 and Theorem 3.2 are the same if X and Y are metric spaces, and Theorem 3.2 is the same as the ϵ-δ definition of continuity.)

Unfortunately, even though Definition 3.3 avoids all mention of ϵ and δ, even if X and Y are metric spaces, it still retains most of the subtlety of ϵ-δ. It turns out that there is yet another way to define continuity which is equivalent to Definition 3.3 but is much easier to remember and often easier to use.

3.4. Theorem. Let X and Y be topological spaces and let $f: X \rightarrow Y$. Then

f is continuous on X if and only if whenever V is an open subset of Y, then $f^{-1}(V)$ is open in X.

[*Hint for proof:* For the "if" part, let x_0 be any point of X and let V be any open subset in Y containing $f(x_0)$; use what is given to find a U to satisfy Definition 3.3. For the "only if" part, let V be open in Y and let x_0 be an arbitrary point in $f^{-1}(V)$; use what is given to find an open subset U of X that contains x_0 and is contained in $f^{-1}(V)$, and then deduce that $f^{-1}(V)$ is open.]

Thus a function is continuous if and only if its *inverse* preserves open sets. Note that this new definition says nothing about continuity at a particular point, but goes directly to a "global" definition of continuity on the whole domain, unlike our previous definitions. Before now we always characterized continuity first at a point, and then said that a function is continuous on its whole domain if it is continuous at every point in the domain. Theorem 3.4 shows that the two approaches to global continuity are the same. However, if you are interested in continuity at a particular point, you will have to use one of our previous definitions; the particular problem at hand usually suggests which definition to use if you are interested in continuity on the whole domain.

3.5. Exercises.

1) Consider $(\mathbf{R}, \mathcal{I})$, $(\mathbf{R}, \mathfrak{D})$, $(\mathbf{R}, \mathfrak{U})$, $(\mathbf{R}, \mathcal{S})$ and $(\mathbf{R}, \mathfrak{F})$, the real line with the indiscrete topology, the discrete topology, the usual metric topology, the Sorgenfrey topology and the finite-complement topology, respectively. Let $f : \mathbf{R} \to \mathbf{R}$ be the identity function defined by $f(r) = r$ for all real numbers r. Determine *all* possible choices for $\mathfrak{I}_1$ and $\mathfrak{I}_2$ from $\mathcal{I}, \mathfrak{D}, \mathfrak{U}, \mathcal{S}, \mathfrak{F}$ so that $f : (\mathbf{R}, \mathfrak{I}_1) \to (\mathbf{R}, \mathfrak{I}_2)$ is continuous. (Use whichever definition of continuity that you think is most convenient as you test each possibility; one may be easier than another for some cases but not for others.)

2) Let X be a set and let $\mathfrak{I}$ be *any* topology on X. There is a topology $\mathfrak{I}'$ that can be assigned to X so that the identity function from $(X, \mathfrak{I}')$ to $(X, \mathfrak{I})$ is always continuous no matter what $\mathfrak{I}$ is. What is it? [*Hint:* Look back at the examples at the beginning of this chapter.]

3) Let X be any set and let $\mathfrak{I}$ be any topology on X. There is a topology $\mathfrak{I}'$ that can be given to X so that the identity function from $(X, \mathfrak{I})$ to $(X, \mathfrak{I}')$ is always continuous no matter what $\mathfrak{I}$ is. What is it? [*Hint:* Look back at the examples at the beginning of this chapter.]

4) Since a function is continuous if and only if its inverse preserves open sets, you might suspect that there is some relation between continuity and a function itself preserving open sets. Such a suspicion would be wrong, though, because there is no relation whatsoever.

A function that preserves open sets is called an **open function.** More precisely, a function $f: X \to Y$ is an open function if whenever U is open in X, then $f(U)$ is open in Y.

a) Give an example of a continuous function that is not open. [*Hint:* See Problem 2 above.]

b) Give an example of an open function that is not continuous. [*Hint:* See Problem 3 above.]

4. NEIGHBORHOODS

Most of our work so far in this chapter has been to generalize the behavior of open sets in some familiar topological spaces. It will turn out that such generalizations will have far-reaching consequences, and will enable us to discover many important and useful results, both about familiar spaces and about non-familiar spaces that occur often in practice. First, though, we want to "generalize the generalization," and obtain a convenient alternative to the term "open set."

Recall that a sequence in **R** converges to a point if it gets close to the point and stays close as n goes to infinity, and we have made the idea of "close" precise in terms of r-balls: a sequence $\{x_n : n \in \mathbf{Z}^+\} \subset \mathbf{R}$ converges to $x \in \mathbf{R}$ if for every positive number r (no matter how small), x_n is ultimately in $S_r(X)$. We might say that x is a "neighbor" of the sequence, then, because x lives close to the sequence, and instead of talking in terms of r-balls, we might change the terminology to suggest this "neighborliness" and call an r-ball an *r-neighborhood.* Usually, we drop the r and say something like this: a sequence $\{x_n : n \in \mathbf{Z}^+\} \subset \mathbf{R}$ converges to x if x_n is ultimately in every neighborhood of x (where a neighborhood of x is simply any set that x belongs to (open or not) that contains an r-ball centered at x for some $r > 0$).

Since open sets generalize r-balls, we can generalize this new idea of neighborhood as follows.

4.1. Definition. Let X be a topological space and let $x \in X$. A set $N \subset X$ is a **neighborhood** of x if there is an open set $U \subset X$ such that $x \in U \subset N$.

Thus N is a neighborhood of a point x if N contains an open set that contains x. Note that N *itself need not be open!* Also, if N is a neighborhood of x, then (obviously) $x \in N$. (It may be obvious, but it is very important to be aware of the fact that a point is an element of any neighborhood of it.)

4.2. Exercises.

1) Consider the spaces $(\mathbf{R}, \mathcal{J})$, $(\mathbf{R}, \mathfrak{D})$, $(\mathbf{R}, \mathfrak{U})$, $(\mathbf{R}, \mathcal{S})$ and $(\mathbf{R}, \mathcal{F})$. ($\mathfrak{U}$ is the usual topology on **R** and $\mathcal{F}$ is the finite complement topology on **R**.) In which of these spaces is:

a) $(0, 2)$ a neighborhood of 1?

  b) [0, 2] a neighborhood of 1?

  c) [0, 2] a neighborhood of 0?

  d) {0} a neighborhood of 0?

2) In the plane with its usual topology, is the unit square,

$$\{(x, y):0 \le x \le 1, 0 \le y \le 1\}$$

a neighborhood of any point in it? Which points?

As the exercises show (and as we have already said) a neighborhood of a point need *not* be an open set. But if a set U *is* open and if $x \in U$, then U is a neighborhood of x. In fact, we can say more.

4.3. Theorem. Let X be a topological space. Then $U \subset X$ is open if and only if U is a neighborhood of each point $x \in U$.

Thus a set is open if and only if it is a neighborhood of each of its points. But we emphasize again that a neighborhood need not be open.

We can extend our definition of sequences to arbitrary topological spaces and define convergence in terms of neighborhoods. (This is a direct generalization of what we did with metric spaces.)

4.4. Definition. Let $\{x_n:n \in \mathbf{Z}^+\}$ be a sequence of points of a topological space X. Then this sequence **converges** to $x \in X$, written $x_n \to x$ or $\lim_{n\to\infty} x_n = x$, if x_n is ultimately in every neighborhood of x.

Using the general definition of convergence of sequences, we can now get two long promised examples, namely, a space in which a sequence can converge to more than one point, and a non-trivial non-metric space.

4.5. Exercises.

1) Let X be the real line with the indiscrete topology and consider the sequence $\{1/n:n \in \mathbf{Z}^+\}$. Prove that if r is *any* point of X, then this sequence converges to r.

    Thus in this space, this sequence converges not only to more than one point but to *every* point in the space—at the same time! But the indiscrete topology on the line is a pretty trivial space because the whole line and the empty set are the only open sets, so there are only two sets in the topology. In the next exercise we get the same result for a non-trivial space.

2) Let $X = \mathbf{R}$ with the finite-complement topology. Prove that $\{1/n:n \in \mathbf{Z}^+\}$ converges to every point in the space.

    Since, in a metric space, if a sequence converges, then it converges to only one point (Theorem 5.4 in Chapter 3), we can now say that the real line with the finite complement topology is *not* a metric space.

3) Recall that a topological space X is a Hausdorff space if distinct

points of X are contained in disjoint open sets (Problem 6 in Exercises 2.6).

a) Prove that a space is a Hausdorff space if and only if distinct points are contained in disjoint neighborhoods.

b) Prove that every metric space is a Hausdorff space.

c) The converse to (b) is false, as we will see in Problem 4 below. First, though, let us observe that the example of a non-metric space that we have from Problem 2 above is also not a Hausdorff space. Prove that in a Hausdorff space, if a sequence converges, then it converges to exactly one point. Deduce that the real line with the finite complement topology is not a Hausdorff space (see Problem 2 above).

4) (This exercise requires a knowledge of ordinal numbers.) From Problems 2 and 3 above, we have an example of a non-Hausdorff space which is not a metric space, which should not be too surprising because we showed (Problem 3(b) above) that a space *must* be a Hausdorff space in order to be a metric space. In this problem, we will get an example of a non-metric space which *is* a Hausdorff space. Thus even though every metric space is a Hausdorff space, there exist Hausdorff spaces which are not metric.

Consider $X = [0, \Omega]$ with the order topology as defined in Section 4 of Chapter 3. We proved (Theorem 5.5 of Chapter 3) that in a metric space, a set S is closed if and only if whenever a sequence of points of S converges to a point x in the space, then $x \in S$.

a) Prove, using the definition of closed set, that $S = [0, \Omega)$ is not a closed subset of $X = [0, \Omega]$ with the order topology.

b) Prove that if a sequence of points of $S = [0, \Omega)$ converges to a point $x \in [0, \Omega]$, then $x \in S$.

c) Deduce that $X = [0, \Omega]$ with the order topology is *not* a metric space, but

d) Prove that $X = [0, \Omega]$ with the order topology *is* a Hausdorff space.

Obviously the point $\Omega \in [0, \Omega]$ is what seems to stop this space from being a metric space. But even if we leave Ω out and consider only $[0, \Omega)$ with the order topology, this space is still not a metric space. However, the proof of this fact is beyond the scope of this book.

We can restate the definition of continuity at a point (Definition 3.3) in terms of neighborhoods, as the following theorem shows.

4.6. Theorem. Let X and Y be topological spaces and let $f: X \to Y$. Then the function f is continuous at a point $x_0 \in X$ if and only if for every neighborhood N_2 of $f(x_0)$ in Y, there is a neighborhood N_1 of x_0 in X such that $f(N_1) \subset N_2$.

Finally, we can reformulate the description of a topological n-manifold in terms of neighborhoods by saying that each point in a topological n-manifold has a neighborhood which is homeomorphic to a neighborhood in E^n or to a neighborhood in $\{(x_1, x_2, \cdots, x_n) : x_i \in \mathbf{R} \text{ and } x_1 \geq 0\}$.

5. CLOSED SETS AND CLOSURE

For convenience, we repeat the definition of closed set here.

5.1. Definition. A subset F of a topological space X is **closed** if its complement $(X - F)$ is open in X.

Thus a set is closed if and only if its complement is open. You should expect the following theorem.

5.2. Theorem. Let X be a topological space. Then
1) X and $\emptyset$ are closed.
2) The intersection of an arbitrary collection of closed sets is closed.
3) The union of a finite collection of closed sets is closed.

Closed sets are just as important in topology as open sets are. Indeed we could have defined a topology on a set by specifying the collection of closed sets (subject to the properties in Theorem 5.2), and then we could have defined a set to be open if its complement is closed—just the opposite of the approach that we took. The important thing is that it would not have made any difference.

5.3. Exercises.
1) Consider again the spaces $(\mathbf{R}, \mathcal{I})$, $(\mathbf{R}, \mathcal{D})$, $(\mathbf{R}, \mathcal{U})$, $(\mathbf{R}, \mathcal{S})$, and $(\mathbf{R}\ \mathcal{F})$, the real line with the indiscrete topology, the discrete topology, the usual metric topology, the Sorgenfrey topology, and the finite-complement topology, respectively. In which of these spaces is:
   a) $[0, 1]$ closed?
   b) $(0, 1)$ closed?
   c) $[0, 1)$ closed?
   d) $(0, 1]$ closed?
   e) $\{0\}$ closed?
2) Recall that a space is a T_1-space if for every pair of distinct points x and y in the space, there is an open set U such that $x \in U$ and $y \notin U$, and there is an open set V such that $y \in V$ and $x \notin V$. (See Problem 6 in Exercises 2.6.)

a) Restate the definition of a T_1-space in terms of neighborhoods.

b) Prove that a space is a T_1-space if and only if for each point p in the space, the singleton set $\{p\}$ is a closed set.

Recall the very handy definition of continuity that we got in Section 3 (Theorem 3.4), namely, that a function between two topological spaces is continuous if and only if its inverse preserves open sets. There is a similar definition in terms of *closed* sets, as is stated in the following theorem.

5.4. Theorem. Let X and Y be topological spaces and let $f: X \to Y$. Then f is continuous (on X) if whenever F is a closed set in Y, then $f^{-1}(F)$ is closed in X.

Thus a function is continuous if and only if its inverse preserves *closed* sets, and a function is continuous if and only if its inverse preserves *open* sets. Both of these statements rank among the most important ones in point-set topology. Sometimes one of them is easier to use than the other, and you should remember them both.

5.5. Exercises.

As with continuity and open sets, you might suspect that since a function is continuous if and only if its inverse preserves closed sets, then there should be some relation between continuity and the function itself preserving closed sets. (See Problem 4 in Exercises 3.5.) As with open sets, such a suspicion would be false because there is no relation whatsoever.

A function $f: X \to Y$ is a **closed function** if whenever F is closed in X then $f(F)$ is closed in Y.

1) Give an example of a closed function that is not continuous. [*Hint*: Consider **R** with the usual topology and with the discrete topology.]

2) Give an example of a continuous function that is not closed. [*Hint*: Consider **R** with the usual topology and with the indiscrete topology.]

For any subset A of a topological space X (whether A is open or closed or neither or both), the smallest closed set that contains A is called the *closure* of A and is defined as follows:

5.6. Definition. Let X be a topological space and let $A \subset X$. Then the **closure** of A (relative to the topology on X) is denoted by $\bar{A}$ and is defined as follows:

$$\bar{A} = \bigcap \{F : F \text{ is closed in } X \text{ and } F \supset A\}.$$

Thus the closure of a set A in X is the intersection of the collection of *all* closed sets in X that contain A. The first thing that we should do with the idea of closure is to justify the name and to show that the description of it as the smallest closed set that contains A is right.

5.7. Theorem. Let X be a topological space and let $A \subset X$. Then
1) $\bar{A}$ is a closed set.
2) $A \subset \bar{A}$.
3) $\bar{A}$ is the smallest closed set that contains A, in the sense that if S is closed and $S \supset A$, then $S \supset \bar{A}$ also.

The definition of closure is sometimes difficult to use, and it is often easier to use the following theorem.

5.8. Theorem. Let X be a topological space and let $A \subset X$. Then if $x \in X$, $x \in \bar{A}$ if and only if $N \cap A \neq \emptyset$ for all neighborhoods N of x.

Thus a point x belongs to $\bar{A}$ if and only if every neighborhood of x meets A.

5.9. Exercises.
1) In the spaces on the real line obtained by giving it the indiscrete topology, the discrete topology, the usual metric topology, the Sorgenfrey topology, and the finite-complement topology, what is $\bar{A}$ if A is
   a) $(0, 1)$.
   b) $[0, 1]$.
   c) $[0, 1)$.
   d) $(0, 1]$.
   e) $\{0\}$.
   f) **Q**, the set of rational numbers.
   g) $\{1/n : n \in \mathbf{Z}^+\}$.
   h) $\emptyset$.
2) Instead of giving you a specific problem, this exercise asks you to formulate the problem (or problems) and then to solve them. (This is more like it is in real life.) Answer the following questions as completely as you can.
   a) How does closure behave with respect to unions?
   b) How does closure behave with respect to intersections?
   c) How does closure behave with respect to complementation?
   d) How does closure behave with respect to Cartesian products?
   e) How do functions affect closure?

One reason why the idea of closure is important is that it gives us yet another way to look at continuity, and the more we know about continuity, the better.

5.10. Theorem. Let X and Y be topological spaces and let $f: X \to Y$. Then
1) f is continuous if and only if whenever $A \subset X$, then $f(\bar{A}) \subset \overline{f(A)}$.
2) f is continuous if and only if whenever $B \subset Y$, then $\overline{f^{-1}(B)} \subset f^{-1}(\bar{B})$.

5.11. Exercises.
1) Show that for a subset A of a topological space, $A = \bar{A}$ if and only if A is closed.
2) We have shown that when A is a subspace of a topological space X, then a point $x \in X$ is in $\bar{A}$ if and only if every neighborhood of x meets A. When every neighborhood of x meets A in at least one point other than x itself, then x is certainly in $\bar{A}$, but this property is more than is needed for x to be in $\bar{A}$.

   We say that $x \in X$ is a **cluster point** of $A \subset X$ if every neighborhood of x meets A in at least one point other than x itself. The set of all cluster points of A is called the **derived set** of A and is denoted by A'. Thus for $x \in X$, $x \in A'$ if and only if for every neighborhood N of x, $N \cap A - \{x\} \neq \emptyset$.

   a) Let **R** have its usual topology, $a, b \in \mathbf{R}$ with $a < b$. What is $(a, b)'$? $[a, b)'$? $[a, b]'$? $\{a\}'$?

   b) Show that if X is any topological space, then $\bar{A} = A \cup A'$. Is $A' = \bar{A} - A$?

   c) Let $X = \{a, b, c, d\}$ with the topology $\mathfrak{I} = \{\emptyset, X, \{a\}, \{a, b\}, \{c, d\}, \{a, c, d\}\}$. What are $\{p\}'$ and $\overline{\{p\}}$ when $p = a, b, c,$ or d? What is $\{a, d\}'$? $\overline{\{a, d\}}$? What is $\{b, d\}'$? $\overline{\{b, d\}}$?

3) Analogous to the closure of a set A being the smallest closed set that contains A, we have the following. The largest open set that is contained in a subset A of a topological space X is called the **interior** of A, is denoted by A°, and is defined by
$$A^\circ = \bigcup \{U \subset X : U \text{ is open and } U \subset A\}.$$
   a) Show that A° is open.
   b) Show that A° is the largest open set contained in A by showing that $A^\circ \subset A$ and if $U \subset X$ is open and $U \subset A$, then $U \subset A^\circ$.
   c) Show that $A = A^\circ$ if and only if A is open.

6. BASIS FOR A TOPOLOGY

As we observed during our discussion of the usual metric topology on the real line, an open set need not be an open interval. (For example, the union of two non-empty disjoint open intervals is not an open interval, but it is an open set.) However, arbitrary open sets were defined in terms of r-balls, and, in the real line, an r-ball is an open interval, so a set is open in R with the usual topology if and only if it contains an open interval around each of its points. Thus open intervals *generate* the topology on the real line, in the sense that an arbitrary open set is the union of open intervals, as we prove in the following theorem. (This theorem was discussed in the section on the order topology; prove it again even if you proved it there.)

6.1. Theorem. A subset of the real line with the usual topology is open if and only if it is the union of a collection of open intervals.

A similar theorem is satisfied by any metric space, and the proof is almost identical to the proof of Theorem 6.1.

6.2. Theorem. A subset of a metric space (X, d) is open if and only if it is the union of a collection of r-balls.

Thus arbitrary open sets in a metric space can be obtained by taking unions of r-balls. Because of this, we say that the collection of all r-balls in a metric space is a *basis* for the metric topology. In general,

6.3. Definition. Let $(X, \mathfrak{I})$ be a topological space. A collection $\mathfrak{B}$ of subsets of X is a **basis** for the topology $\mathfrak{I}$ if
1) Every member of $\mathfrak{B}$ is open in $(X, \mathfrak{I})$ (i.e., $\mathfrak{B} \subset \mathfrak{I}$) and
2) Each open subset of $(X, \mathfrak{I})$ (i.e., each member of $\mathfrak{I}$) is the union of a collection of sets in $\mathfrak{B}$.
The members of $\mathfrak{B}$ are called **basic open sets** in $(X, \mathfrak{I})$.

6.4. Exercises.
1) Give a basis for **R** when it has the
   a) usual metric topology.
   b) the indiscrete topology.
   c) the discrete topology.
   d) the Sorgenfrey topology.
   e) the finite-complement topology.
2) a) What is a basis for the order topology on $\mathbf{Z}^+$? On **R**?
   b) (This exercise requires a knowledge of ordinal numbers.) What is a basis for the order topology on $[0, \Omega)$?
   c) Answer all three questions in (a) and (b) by giving a basis for the order topology on an arbitrary totally ordered set X.
3) An r-ball in the plane is an open disk (a disk without its edge). According to Theorem 6.2 and the definition of a basis for a topology, the collection of all of these open disks is a basis for the usual metric topology on the plane. But a given topology can have more than one basis, as you may have observed in doing Problem 1 above.
   a) Show that the collection of open squares (squares of arbitrary size without their edges) is a basis for the usual metric topology on the plane.
   b) Can you think of any other bases for the usual metric topology on the plane?

As we saw in Exercises 6.4, a given topology can have more than one basis. In a metric space, there is one particular basis for the metric topology which is especially important.

6.5. Theorem. Let (X, d) be a metric space. Then the collection $\mathcal{B} = \{S_{1/n}(x) : n \in \mathbf{Z}^+ \text{ and } x \in X\}$ is a basis for the metric topology on (X, d).

Thus the collection of $1/n$-balls centered at each point in the space is a basis for the metric topology on (X, d).

6.6. Definition. Let $(X, \mathfrak{I})$ be a topological space, and let $x \in X$. A collection of sets $\mathcal{B}_x$ is a **local basis** at x if

1) each member of $\mathcal{B}_x$ is open(i.e., $\mathcal{B}_x \subset \mathfrak{I}$), and
2) For any open set U containing x, there exists a $B \in \mathcal{B}_x$ such that $x \in B \subset U$.

Obviously, for a given point x in a metric space (X, d), the collection of $1/n$-balls centered at x is a local basis at x. Furthermore, this local basis is a *countable* collection of sets, and this is why this particular local basis is so important. Recall that we proved (Theorem 5.5 in Chapter 3) that in a metric space, a set F is closed if and only if whenever a sequence of points of F converges to a point x in the space, then $x \in F$. The proof of the "only if" part of this theorem involves constructing a sequence of points of F that converges to x, and this construction rests on the fact that any open set containing x contains a $1/n$-ball centered at x, for some $n \in \mathbf{Z}^+$. The fact that the collection of $1/n$-balls centered at x is a local basis at x is precisely what enables us to obtain such a $1/n$-ball contained in any given open set, and allows us to prove the theorem.

A similar theorem holds whenever a space has a countable local basis at each point.

6.7. Definition. A topological space X is called **1st countable** or is said to satisfy the **first axiom of countability** if there is a countable local basis at each point of the space.

6.8. Theorem. Let X be a 1st countable topological space. Then a subset $F \subset X$ is closed if and only if whenever $\{x_n : n \in \mathbf{Z}^+\}$ is a sequence of points of F that converges to $x \in X$, then $x \in F$.

Because of this theorem, we say that the closed sets in a 1st countable space (and, in particular, in a metric space) can be completely characterized using sequences. In other words, to decide if a set is closed in a 1st countable space, we need only consider (countable) sequences of points.

6.9. Exercises.

1) Give an example of a topological space which is not 1st countable.

[*Hint:* Consider the order topology on an appropriate set, *or* consider the finite-complement topology on an uncountable set.]

2) Let $(X, \mathfrak{I})$ be a topological space and let $\mathfrak{B}_x$ be a local basis at $x \in X$. Can $\cap \mathfrak{B}_x = \{x\}$? Must $\cap \mathfrak{B}_x = \{x\}$?

3) a) Let $(X, \mathfrak{I})$ be a topological space and, for each point $x \in X$, let $\mathfrak{B}_x$ be a local basis at x. Show that $\mathfrak{B} = \cup\{\mathfrak{B}_x : x \in X\}$ is a basis for the topology on X.

   b) Let $(X, \mathfrak{I})$ be a topological space and let $\mathfrak{B}$ be a basis for $\mathfrak{I}$. Show that for each point $x \in X$, the collection

$$\mathfrak{B}_x = \{B \in \mathfrak{B} : x \in B\}$$

is a local basis at x.

Using the idea of a basis for a topology can simplify checking continuity of a function. For example, we know that a function between two topological spaces is continuous if and only if its inverse preserves open sets. Thus, to check continuity of a given function according to this definition, we would have to show that for any open set whatsoever in the range of the function, the inverse of that set is open in the domain. Working with arbitrary open sets, though, can be cumbersome and even difficult and confusing in some cases. Using the idea of a basis for a topology, we can eliminate the need to deal with arbitrary open sets in the range, as the following theorem shows.

6.10. Theorem. Let $(X, \mathfrak{I}_1)$ and $(Y, \mathfrak{I}_2)$ be topological spaces, let $\mathfrak{B}$ be a basis for the topology $\mathfrak{I}_2$ on Y, and let $f: X \to Y$. Then f is continuous on X if and only if for any $B \in \mathfrak{B}$, $f^{-1}(B)$ is open in $(X, \mathfrak{I}_1)$.

Thus a function is continuous if and only if the inverse of a *basic* open set is open. This result says, for example, that to check continuity of a function whose range is **R** (with the usual topology), we need only check that the inverse of an open interval is an open set. Notice that we do *not* say that the inverse of an open interval is an open interval, nor can we say in general that the inverse of a basic open set is a basic open set, as the following exercises show.

6.11. Exercises.

Let $\mathbf{R}^+ \cup \{0\}$ be the set of non-negative real numbers and give it the usual metric topology induced by the absolute value metric.

1) Describe a basis for this topology.

2) Give an example of a continuous function $f: \mathbf{R} \to \mathbf{R}^+ \cup \{0\}$ (where **R** has the usual metric topology) such that there is a basic open set in $\mathbf{R}^+ \cup \{0\}$ whose inverse is not a basic open set in **R**.

The convenience of using basic open sets rather than arbitrary open sets can be combined with the convenience of using neighborhoods. Recall that a neighborhood of a point is simply a set that contains an open set that contains the point, but that a neighborhood itself need not be open. We can modify this definition of neighborhood slightly.

6.12. Theorem. A subset N of a topological space X with basis $\mathcal{B}$ is a neighborhood of a point $x \in X$ if and only if there is a basic open set $B \in \mathcal{B}$ such that $x \in B \subset N$.

Besides being easier to use in many cases, the idea of a neighborhood of a point being a set that contains a basic open set that contains the point is what most people mean when they talk about neighborhoods. Sometimes a basic open set is called a **basic open neighborhood,** and we will often use this terminology. But remember that in general, a neighborhood need not be open. (It complicates things needlessly to require that neighborhoods be open because it is usually not important whether a neighborhood is open or not.)

6.13. Definition. Let X be a topological space, let $x \in X$ and let $\mathcal{B}_x$ be a local basis at x. A **local neighborhood basis** at x is a collection $\mathfrak{N}$ of subsets of X, each of which contains x, such that for every $N \in \mathfrak{N}$, there is a $B \in \mathcal{B}_x$ such that $x \in B \subset N$.

6.14. Exercises.
   Show that the collections of sets $\{[a, b]:a < 0 < b\}$, $\{[a, b):a < 0 < b\}$, $\{(a, b]:a < 0 < b\}$, and $\{(a, b):a < 0 < b\}$ are all local neighborhood bases at 0 in the usual topology on the real line. Which of these collections is a local basis at 0 in the usual topology on **R**?

7. TOPOLOGY GENERATED BY A BASIS

When we defined the usual metric topology on the real line, we said that a set is open if and only if it contains an r-ball around each of its points, and then we proved that a set is open in this space if and only if it is the union of a collection of r-balls (actually a collection of open intervals in the case of the real line). Because of this last result, we say that the usual metric topology on the real line is the topology generated by the collection of all r-balls, in the following sense.

7.1. Definition. Let $\mathcal{B}$ be a collection of subsets of a set X, and let $\mathcal{B}*$ be the collection of all sets which are unions of members of $\mathcal{B}$, together with the empty set. If $\mathcal{B}*$ is a topology on X (i.e., if $\mathcal{B}*$ satisfies (1), (2), and (3) of Definition 1.1), then $\mathcal{B}*$ is called the **topology generated by** $\mathcal{B}$, and $\mathcal{B}$ is a basis for the topology $\mathcal{B}*$.

Thus the usual metric topology on the real line is $\mathcal{B}*$ when $\mathcal{B}$ is the collection of all open intervals.

Definition 7.1 raises two important questions. First, what conditions on $\mathcal{B}$ do we need to guarantee that $\mathcal{B}*$ is a topology on X? Second, suppose that we already have a topology on X (as we do on the real line, for example), and that $\mathcal{B}$ is such that $\mathcal{B}*$ is also a topology on X. How can we be sure that $\mathcal{B}*$ is the same as the topology that we already have? The following exercises show that both of these questions need to be asked.

7.2. Exercises.
1) Let $\mathcal{B}$ be the collection of all closed intervals on the real line. Show that $\mathcal{B}*$, the collection of all possible unions of elements of $\mathcal{B}$, together with the empty set, is *not* a topology on **R**.
2) Let **R** have the usual topology and consider the collection $\mathcal{B}$ of subsets of **R** defined by

$$\mathcal{B} = \{\{x\} : x \in \mathbf{R}\}$$

(so $\mathcal{B}$ is the collection of all singleton subsets of **R**).
   a) Show that $\mathcal{B}*$ *is* a topology on the set **R**, but
   b) Show that $\mathcal{B}*$ is not the usual topology on **R**.

To tell when $\mathcal{B}*$ is a topology on X we have the following theorem.

7.3. Theorem. Let X be a set and let $\mathcal{B}$ be a collection of subsets of X. Let $\mathcal{B}*$ be the collection of all possible unions of members of $\mathcal{B}$, together with the empty set. Then $\mathcal{B}*$ is a topology on X if and only if
1) $\cup\mathcal{B} = X$, and
2) For B_1 and B_2 in $\mathcal{B}$ and any point $x \in B_1 \cap B_2$, there exists a set $B_3 \in \mathcal{B}$ such that $x \in B_3 \subset B_1 \cap B_2$.

When $\mathcal{B}$ satisfies (1) and (2), we will say that it is a **basis for a topology** on X.

The first condition in Theorem 7.3 merely ensures that X itself is open in the topology $\mathcal{B}*$. That the empty set and arbitrary unions of members of $\mathcal{B}*$ belong to $\mathcal{B}*$ is automatic from the way $\mathcal{B}*$ is defined. The only thing left for $\mathcal{B}*$ to be a topology is that finite intersections of members of $\mathcal{B}*$ belong to $\mathcal{B}*$, and this is taken care of by (2). Notice that in (2) we do *not* require that $B_1 \cap B_2$ itself belong to $\mathcal{B}$, but only that it contain a member of $\mathcal{B}$ containing any given point of $B_1 \cap B_2$. The set $B_1 \cap B_2$ does, however, belong to $\mathcal{B}*$ when $\mathcal{B}$ satisfies (2).

Thus a collection of subsets of a set X is a basis for a topology on X if it satisfies certain properties (namely, (1) and (2) of Theorem 7.3). Note the distinction here relative to our previous discussion of basis. Here we have a collection of sets, and we use this collection to generate a topology, without knowing anything about the topology before we start. Once we get the topology, then the collection of sets that we started with is a basis for it in our previous sense.

To know if ℬ* is the same as some preassigned topology, we have the following.

7.4. Theorem. Let $(X, \mathfrak{I})$ be a topological space, let ℬ be a basis for a topology on X, and let ℬ* denote the topology for which ℬ is a basis (i.e., ℬ generates ℬ*). Then ℬ* = $\mathfrak{I}$ if and only if

1) For every set $B \in$ ℬ and every point $x \in B$, there is a $U \in \mathfrak{I}$ such that $x \in U \subset B$, and
2) For every set $U \in \mathfrak{I}$ and every point $x \in U$, there is a $B \in$ ℬ such that $x \in B \subset U$.

7.5. Exercises.

Are the following statements true or false?

1) A topology on a set can have more than one basis.
2) A basis for a topology on a set can generate more than one topology.

Recall that a basic open set is often called a basic open neighborhood. The following theorem is actually just a restatement of previous discussion, but it gives the terminology that we will use in practice when dealing with generating a topology from a basis. You should be absolutely sure that you understand what it says and how to use it.

7.6. Theorem. Let ℬ be a basis for the topology $\mathfrak{I}$ on a set X (i.e., ℬ generates $\mathfrak{I}$ so $\mathfrak{I}$ = ℬ*). A subset $U \subset X$ is open in $\mathfrak{I}$ = ℬ* if and only if for any $x \in U$ there exists a basic open neighborhood $N \in$ ℬ of x such that $x \in N \subset U$.

Thus a subset U of X is open in the topology generated by a basis ℬ if and only if U contains a basic open neighborhood (i.e., a member of ℬ) of each of its points.

7.7. Exercises.

1) We have already verified most of the facts in this exercise, but metric spaces are so important that we collect them here for review and future reference.

    Let d be a metric on a set X (so $d: X \times X \to \mathbf{R}$ is a non-negative function which is anti-symmetric, transitive, and satisfies the triangle inequality. (See Definition 2.1 in Chapter 3 if you don't remember what these words mean.)) Then

    a) the collection of all r-balls,

    $$\text{ℬ} = \{S_r(x) : r \in \mathbf{R}, r > 0 \text{ and } x \in X\},$$

is a basis for a topology on X, the metric topology induced by d.

b) For a given $x \in X$, the collection of all r-balls centered at x,

$$\mathcal{B}_x = \{S_r(x) : r \in \mathbf{R} \text{ and } r > 0\},$$

is a local basis at x in the metric topology induced by d.

c) The collection $\{S_{1/n}(x) : n \in \mathbf{Z}^+\}$ is a local basis at x in the metric topology induced by d.

2) Let $X = [-1, 1]$ and let

$$\mathcal{B} = \{[-1, b), (a, 1] : -1 < a, b < 1\} \cup \{0\}.$$

Then $\mathcal{B}$ is a basis for a topology on $[-1, 1]$ which is different from any that we have considered previously. In this topology, the singleton set $\{0\}$ is an open set, but no other singleton set is open.

3) Let $g([0, 1])$ denote the collection of all functions from $[0, 1]$ into $\mathbf{R}$ such that $\int_0^1 |f(x)| \, dx$ exists.

a) We can make this collection of functions into a topological space in which the *points* of the space are *functions* by specifying a basis for the topology. For $f \in g([0, 1])$, let

$$\mathcal{B}_f = \{B_r(f) : r \in \mathbf{R} \text{ and } r > 0\}$$

where

$$B_r(f) = \left\{g \in ([0, 1]) : \int_0^1 |f(x) - g(x)| \, dx < r\right\}.$$

Let $\mathcal{B} = \bigcup_{f \in g([0,1])} \mathcal{B}_f$. Show that $\mathcal{B}$ is a basis for a topology on $g([0, 1])$.

b) Show that the "distance function" implicitly defined in (a), namely

$$d(f, g) = \int_0^1 |f(x) - g(x)| \, dx$$

is *not* a metric on $g([0, 1])$. [*Hint:* If $f(x) = 1$ for $0 < x \le 1$ and $f(0) = 2$, does $\int_0^1 |f(x)| \, dx$ exist?]

4) Let $\mathcal{C}([0, 1])$ denote the collection of all continuous real-valued functions defined on $[0, 1]$. Define a "distance function" on $\mathcal{C}([0, 1])$ by

$$d(f, g) = \int_0^1 |f(x) - g(x)| \, dx.$$

Is this distance function a metric? Describe the topology generated by this distance function as in (3) above.

5) Even if a distance function is not a metric, it may be close enough to being a metric so that r-balls make sense. What do r-balls look like using the distance function of exercises (3) and (4)?

6) Let $\mathcal{C}([0, 1])$ denote the collection of all continuous real-valued functions defined on $[0, 1]$.

a) For each $f \in \mathcal{C}([0, 1])$, let

$$\mathcal{B}_f = \{ B_r(f) : r \in \mathbf{R} \text{ and } r > 0 \}$$

where

$$B_r(f) = \{ g \in \mathcal{C}([0, 1]) : \max_{0 \leq x \leq 1} |f(x) - g(x)| < r \}$$

(where $\max_{0 \leq x \leq 1} |f(x) - g(x)|$ means the maximum value that $|f(x) - g(x)|$ attains as x ranges over all possible values in $[0, 1]$; we have not proved it yet but this maximum value always exists.)

Let $\mathcal{B} = \bigcup_{f \in \mathcal{C}([0,1])} \mathcal{B}_f$. Show that $\mathcal{B}$ is a basis for a topology on $\mathcal{C}([0, 1])$.

b) The distance function defined on $\mathcal{C}([0, 1])$ by

$$d(f, g) = \max_{0 \leq x \leq 1} |f(x) - g(x)|$$

is actually a metric on $\mathcal{C}([0, 1])$. What do r-balls look like in this metric?

8. SUB-BASIS FOR A TOPOLOGY

One reason why bases are important is that they allow us to reduce the number of open sets that we need to deal with in many cases (in discussing continuity, for example). It turns out that we can reduce the number of open sets even further.

8.1. Definition. Let $(X, \mathfrak{I})$ be a topological space. A collection $\mathcal{S}$ of subsets of X is a **sub-basis** for $\mathfrak{I}$ if the collection of all finite intersections of members of $\mathcal{S}$ is a basis for $\mathfrak{I}$.

The members of $\mathcal{S}$ are called **sub-basic open sets**.

Thus a sub-basis is a collection of sets that generates a basis by taking finite intersections, and the basis then generates the topology by taking arbitrary unions. For example, the collection $\mathcal{S} = \{(-\infty, b), (a, \infty): a, b \in \mathbf{R}\}$ is not a basis for the usual topology on $\mathbf{R}$ (it fails (2) of Theorem 7.3). But this collection of subsets of $\mathbf{R}$ *is* a sub-basis for the usual topology on $\mathbf{R}$ because the collection of all finite intersections of members of $\mathcal{S}$ contains the collection $\mathcal{B} = \{(a, b): a, b \in \mathbf{R}\}$ which, as we know, is a basis for the usual topology on $\mathbf{R}$.

Unlike the situation with bases, *any* collection of subsets of a set X is a sub-basis for a topology on X (although not, of course, necessarily a sub-basis for some *particular* topology that we might already have on X). The proof of this fact should be easy if you understand the definitions of both a sub-basis and a basis.

8.2. Theorem. Let $\mathcal{S}$ be any collection of subsets of a set X. Then $\mathcal{S}$ is a sub-basis for a topology on X.

The relation between sub-bases and continuity is a good one.

8.3. Theorem. Let $(X, \mathcal{T}_1)$ and $(Y, \mathcal{T}_2)$ be topological spaces and let $\mathcal{S}$ be a sub-basis for $\mathcal{T}_2$. Then a function $f: X \to Y$ is continuous on X if and only if $f^{-1}(S)$ is open in $(X, \mathcal{T}_1)$ for all $S \in \mathcal{S}$.

Thus to check continuity, we need only check that the inverse of a *sub-basic* open set is open, which often greatly simplifies the work. Notice that we do not say that $f^{-1}(S)$ is a sub-basic open set or even a basic open set; all we can say is that it is an open set.

Perhaps the most important use of sub-bases occurs when we discuss Cartesian products. Recall that the Euclidean plane is the set $\mathbf{R} \times \mathbf{R}$, and that it is a metric space with the metric topology induced by the distance formula

$$d((x_1, x_2), (y_1, y_2)) = \sqrt{(x_1 - y_1)^2 + (x_2 - y_2)^2}.$$

Thus the topology on the plane is generated by the collection of all open disks (without their edges), because an r-ball in the distance formula metric is such a disk, and the collection of all r-balls is a basis for the metric topology. But as we have seen several times, a topology can have more than one basis, and the collection of all rectangles in the plane (without their edges) is also a basis for the metric topology on the plane. (This is similar to Problem 3 of Exercises 6.4; show it now even if you did the exercise.) A rectangle in the plane, though, is the intersection of two "strips," as the following picture illustrates.

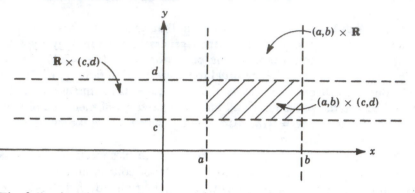

The following theorem should now be easy to prove.

8.4. Theorem. The collection of subsets of the plane

$$\mathbb{S} = \{(a, b) \times \mathbf{R}, \mathbf{R} \times (c, d) : a < b, c < d, a, b, c, d \in \mathbf{R}\}$$

of all "strips" in the plane is a sub-basis for the usual topology on the plane.

The sub-basis of strips in the plane is of great importance in topology because it enables us to generalize the topology on the plane (which is the Cartesian product $\mathbf{R} \times \mathbf{R}$) to other Cartesian products. The terminology is easier if we use the *projection functions*, which are defined as follows.

8.5. Definition. Let X and Y be topological spaces. Then there are two **projection functions** defined on $X \times Y$. These are $\pi_1 : X \times Y \to X$ and $\pi_2 : X \times Y \to Y$, defined by $\pi_1(x, y) = x$ and $\pi_2(x, y) = y$.

The following picture illustrates this definition.

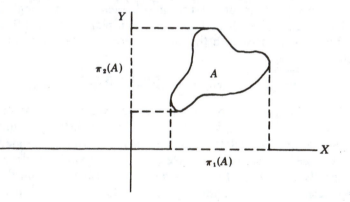

Note that $A \neq \pi_1(A) \times \pi_2(A)$. Indeed, $\pi_1(A) \times \pi_2(A)$ is necessarily a rectangle in the "plane" $X \times Y$, whereas A can be a subset of any shape at all.

8.6. Exercises.
1) A subset A of $\mathbf{R} \times \mathbf{R}$ is *bounded* if there exists an integer n such that $A \subset S_n((0, 0))$. Show that for any bounded subset A of $\mathbf{R} \times \mathbf{R}$, $\pi_1(A) \times \pi_2(A)$ is a rectangle, no matter what the shape of A is.
2) For $a, b \in \mathbf{R}$, show that $\pi_1^{-1}((a, b)) = (a, b) \times \mathbf{R}$, and $\pi_2^{-1}((a, b)) = \mathbf{R} \times (a, b)$.

It is true that even though the projection functions π_1 and π_2 do not preserve the shape of a set in any way, even on $\mathbf{R} \times \mathbf{R}$, nor do their inverses, their inverses *can* be used to describe a sub-basis for the topology on $\mathbf{R} \times \mathbf{R}$. The proof of the following theorem should now be easy, and the theorem itself will be of great importance to us in our discussion of product spaces in general in the next chapter.

8.7. Theorem. The collection of subsets of the plane

$$\mathcal{S} = \{\pi_1^{-1}((a, b)), \pi_2^{-1}((a, b)) : a, b \in \mathbf{R} \text{ and } a < b\}$$

is a sub-basis for the usual topology on the plane.

Also, even though they do not preserve shape, the projection functions *are* continuous. Any discussion of continuity of a function between two spaces requires a topology on each of the spaces, and we have not defined a topology on $X \times Y$ in general. However, we do have a topology on $\mathbf{R} \times \mathbf{R}$, and we can prove continuity of the projection functions in this case.

8.8. Theorem. The projection functions π_1 and π_2 from $\mathbf{R} \times \mathbf{R}$ into $\mathbf{R}$ are continuous when $\mathbf{R} \times \mathbf{R}$ and $\mathbf{R}$ have their usual topologies.

Continuous functions are often called **maps** or **mappings,** and the projection functions are then called **projection maps** or **projection mappings.** We will use this terminology, and justify it in the next chapter by proving that projection functions in general are continuous.

9. SUBSPACES

We have already had occasion to make certain subsets of topological spaces into topological spaces themselves. For example, we said that a closed interval $[a, b] \subset \mathbf{R}$ (with $a < b$) can be given the topology generated by the usual absolute value metric, and a circle (which is a subset of the plane) can be given a topology by using a basis of "curved open intervals." It is natural to ask if there is any relation between the topologies that we get on these subsets and the usual topologies that we already have on the line and the plane. If there is such a relationship, it is *not* the very simple one that a set is open in a subset of a space if and only if it is open in the space itself, because, for example, $[0, \frac{1}{2})$ is open in the topology that we gave to $[0, 1]$ while it is not open in **R** with the usual topology. Similarly, no "curved interval" on the circle is open in the plane.

But there is a relation and it is very simple one, if not quite as simple as we might hope. Consider the following picture.

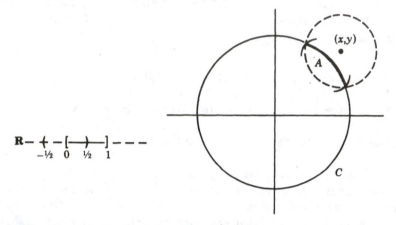

The interval $[0, \frac{1}{2})$ is open in $[0, 1]$ but not in **R**; however $[0, \frac{1}{2}) = (-\frac{1}{2}, \frac{1}{2}) \cap [0, 1]$ and $(-\frac{1}{2}, \frac{1}{2})$ *is* open in **R**. Similarly, the "curved interval" A on the circle C is open in the circle but not in the plane; but $A = S_r((x, y)) \cap C$, and $S_r((x, y))$ *is* open in the plane.

Thus it should seem reasonable to say that a set is open in a subset of a space if it is the intersection of the subset with a set which is open in the space itself. We make this precise as follows.

9.1. Definition. Let X be a topological space and let $A \subset X$. The **relative topology** on A (inherited from the topology on X) is defined by: $U \subset A$ is open in the relative topology on A if and only if there exists an open subset V of X such that $U = V \cap A$.

It should not be too hard to convince yourself that the topologies described above on a closed interval and a circle are indeed the relative topologies inherited from the line and the plane, respectively. The following theorem might make it easier.

9.2. Theorem. Let $(X, \mathfrak{I})$ be a topological space and let $A \subset X$. Then if $\mathfrak{B}$ is a basis for $\mathfrak{I}$ (on X), the collection

$$\mathfrak{B}_A = \{B \cap A : B \in \mathfrak{B}\}$$

is a basis for the relative topology on A.

[*Hint for proof*: Show that a subset of A is open in the relative topology on A if and only if it is open in the topology generated by $\mathfrak{B}_A$. Recall that a set is open in the topology generated by a basis if and only if it contains a basic open neighborhood of each of its points.]

When $A \subset X$ has the relative topology (inherited from the topology on X), we will usually refer to A as a **subspace** of X, rather than as a subset of X.

9.3. Exercises.

1) Show that **R** with its usual topology is a subspace of E^2, the Euclidean plane with its usual topology. [*Hint:* Identify **R** with the x-axis in the plane and see Theorem 9.2.]

2) Describe the topology that $\mathbf{Z}^+$ gets as a subspace of **R**.

3) (This exercise requires a knowledge of ordinal numbers.) Describe the topology that $\mathbf{Z}^+$ gets as a subspace of the space of countable ordinals, $[0, \Omega)$. Thus, combining this result with Problem 2 above, we see that a set can be topologically the same even when it is viewed as a subspace of two very different topological spaces.

4) Show that $\mathbf{Z}^+$ and the set $\{1/n : n \in \mathbf{Z}^+\}$ are topologically the same (i.e., are homeomorphic) when both are viewed as subspaces of **R** with its usual topology. In particular, $\{1/n : n \in \mathbf{Z}^+\}$ is a discrete space, even though there are points in this set that are as close together as we want. Is $\{1/n : n \in \mathbf{Z}^+\} \cup \{0\}$ a discrete subspace of **R**?

5) If $\mathcal{S}$ is a sub-basis for a topology $\mathfrak{I}$ on X and $A \subset X$, is $\{S \cap A : S \in \mathcal{S}\}$ a sub-basis for the relative topology on A?

Closed sets behave very nicely with respect to the relative topology. Specifically,

9.4. Theorem. Let X be a topological space and let $A \subset X$ have the relative topology inherited from the topology on X. A subset $F \subset A$ is closed in the relative topology on A if and only if there exists a closed subset C of X such that $F = C \cap A$.

Thus a subset $A \subset X$ is *open* in the relative topology on A if and only if it is the intersection of A with an *open* subset of X, and a subset $A \subset X$ is *closed* in the relative topology on A if and only if it is the intersection of A with a *closed* subset of X.

The closure of a set in the relative topology is also very well behaved.

For $S \subset X$, we have denoted the closure of S by $\bar{S}$, and we showed that a point $x \in X$ is an element of $\bar{S}$ if and only if every neighborhood of x meets S (Theorem 5.8). If $A \subset X$ has the relative topology and $S \subset A$, then there are two possible meanings for $\bar{S}$: it could mean the closure of S with respect to A, and it could mean the closure of S with respect to X. The two need not be the same! (See Exercise 9.6.) Let us continue to denote the closure of $S \subset X$ with respect to X by $\bar{S}$, but the closure of $S \subset A \subset X$ with respect to A, when A has the relative topology inherited from X, will be denoted by $\bar{S}^A$. Then

9.5. Theorem. Let X be a topological space and let $A \subset X$ have the relative topology inherited from the topology on X. With the above notation, if $S \subset A$, then $\bar{S}^A = \bar{S} \cap A$.

Thus the closure of $S \subset A$ with respect to the relative topology on A inherited from X is simply the interesection of A with the closure of S in X.

9.6. Exercise.

Give an example to show that when A is a subspace of a topological space X and $S \subset A$, then $\bar{S}^A$ need not be the same as $\bar{S}$. [*Hint:* For a point to be in the closure of a set with respect to a topological space, the point must necessarily be an element of the space.]

As we have observed, an open subset of A need not be open in X when $A \subset X$ has the relative topology inherited from X. Similarly, a closed subset of A need not be closed in X. However, A itself may be open or closed in X, and this makes a difference, as the following theorem shows.

9.7. Theorem. Let X be a topological space and let $A \subset X$ have the relative topology inherited from the topology on X. Then
  1) If A is an open subset of X, then $U \subset A$ is open in A if and only if it is open in X.
  2) If A is a closed subset of X, then $F \subset A$ is closed in A if and only if it is closed in X.

When $f: X \to Y$ is continuous and $A \subset X$ is a subspace of X, then we can talk about the continuity of the function f *restricted to the subspace* A.

9.8. Definition. Let X and Y be sets and let $f: X \to Y$. If $A \subset X$, then the **restriction** of f to A, denoted $f \,|\, A$, is the function from A to Y defined by $(f \,|\, A)(x) = f(x)$ for $x \in A$.

Thus $(f \,|\, A)(x)$ is the same as $f(x)$ when $x \in A$, and $(f \,|\, A)(x)$ is simply not defined when $x \notin A$.

9.9. Theorem. Let X and Y be topological spaces, and let $A \subset X$ have the relative topology inherited from X (i.e., A is a subspace of X). If $f: X \to Y$ is a continuous function, then $f \mid A : A \to Y$ is also continuous.

Recall that if $f: X \to Y$ and $g: Y \to Z$ then the *composition* of f and g is the function $g \circ f : X \to Z$ defined by $g \circ f(x) = g(f(x))$.

9.10. Theorem. Let X, Y, and Z be topological spaces and let $f: X \to Y$, $g: Y \to Z$. If f and g are both continuous, then $g \circ f : X \to Z$ is also continuous.

[*Hint for proof:* Since $g: Y \to Z$ is continuous, then $g \mid f(X) : f(X) \to Z$ is also continuous by Theorem 9.9. The "inverse of open sets is open" definition of continuity is probably the easiest one to use here.]

9.11. Exercise.

Give an example to show that the converse of Theorem 9.10 is false. [*Hint:* See Problems 2 and 3 in Exercises 3.5.]

10. CONTINUITY AGAIN

We have seen several equivalent definitions of continuity in this Chapter. For later reference, we collect them here.

10.1. Theorem. Let X and Y be topological spaces and let $f: X \to Y$. Then the following statements are equivalent.

1) The function f is continuous on X.
2) For each point $x \in X$ and each neighborhood U of $f(x)$, there exists a neighborhood V of x such that $f(U) \subset V$.
3) For each open subset U of Y, $f^{-1}(U)$ is open in X.
4) For each closed subset F of Y, $f^{-1}(F)$ is closed in X.
5) For any subset A of X, $f(\bar{A}) \subset \overline{f(A)}$.
6) For any subset B of Y, $\overline{(f^{-1}B)} \subset f^{-1}(\bar{B})$.

chapter five

Product Spaces

Product spaces are of great importance in analysis and topology. In this chapter we will define Cartesian products in general (generalizing the definition given for $X \times Y$), and will define the product topology on these Cartesian products as a generalization of the topology on the Euclidean plane. We will then examine some of the properties of the resultant *product spaces*.

To avoid needless complications, we will assume throughout this Chapter that all sets mentioned (including all topological spaces) are non-empty.

1. THE PRODUCT OF TWO SPACES

For convenience, we repeat some of our previous definitions here.

1.1. Definition. Let X_1 and X_2 be sets. The **Cartesian product** of X_1 and X_2, denoted $X_1 \times X_2$ (read "X_1 cross X_2"), is defined by

$$X_1 \times X_2 = \{(x_1, x_2) : x_1 \in X_1 \text{ and } x_2 \in X_2\}.^*$$

Thus $X_1 \times X_2$ is the set of ordered pairs with first coordinate an element of X_1 and second coordinate an element of X_2. The sets X_1 and X_2 are called the **factors** of the product $X_1 \times X_2$. Two ordered pairs are equal if and only if they are equal coordinatewise. In other words, $(x_1, x_2) = (y_1, y_2)$ if and only if $x_1 = y_1$ and $x_2 = y_2$.

The **projection functions** $\pi_1 : X_1 \times X_2 \to X_1$ and $\pi_2 : X_1 \times X_2 \to X_2$ are defined by $\pi_1(x_1, x_2) = x_1$ and $\pi_2(x_1, x_2) = x_2$.

---

* Note that according to this definition, $X_1 \times X_2 = \emptyset$ if and only if at least one of X_1 and X_2 is empty. But we have made the blanket assumption in this Chapter that all sets are non-empty, so we need not worry about $X_1 \times X_2$ being the empty set.

When X_1 and X_2 are topological spaces, we can define a topology on $X_1 \times X_2$ by using the projection functions, as we did on $\mathbf{R} \times \mathbf{R}$ in Section 8 of Chapter 4. Specifically,

1.2. Definition. Let $(X_1, \mathfrak{I}_1)$ and $(X_2, \mathfrak{I}_2)$ be topological spaces. The **product topology** on $X_1 \times X_2$ is the topology on $X_1 \times X_2$ generated by the sub-basis

$$\{\pi_1^{-1}(U_1), \pi_2^{-1}(U_2) : U_1 \in \mathfrak{I}_1 \text{ and } U_2 \in \mathfrak{I}_2\}.$$

The set $X_1 \times X_2$ with the product topology is the **product space** $X_1 \times X_2$.

As we mentioned before (Section 3 of Chapter 1), it is often very useful to think of $X_1 \times X_2$ as a plane (even though it usually is not), and to draw pictures with X_1 thought of as the x-axis and X_2 the y-axis in this plane. Of course such pictures will not prove anything, but they may help you to see what is going on.

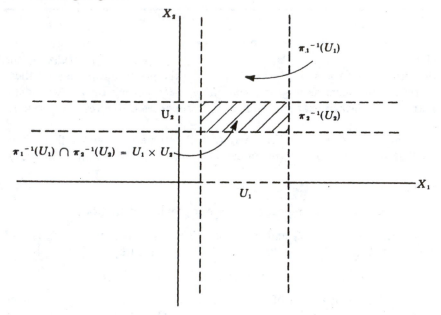

1.3. Exercises.

1) Prove that for $U_1 \subset X_1$, $\pi_1^{-1}(U_1) = U_1 \times X_2$, and for $U_2 \subset X_2$, $\pi_2^{-1}(U_2) = X_1 \times U_2$.

2) Show that for $U_1 \subset X_1$ and $U_2 \subset X_2$, $\pi_1^{-1}(U_1) \cap \pi_2^{-1}(U_2) = U_1 \times U_2$.

Since $\pi_1^{-1}(U_1) = U_1 \times X_2$ and $\pi_2^{-1}(U_2) = X_1 \times U_2$, the product topology on $X_1 \times X_2$ is the topology generated by the sub-basis consisting of all "strips" in the "plane" $X_1 \times X_2$ over open subsets of X_1 and X_2. (See figure on page 90.) In general, $U \times X_2$ is called a **strip over** U, and $X_1 \times V$ is a **strip over** V).

Actually, we need not use all of the open sets in X_1 and X_2 to generate the product topology on $X_1 \times X_2$.

1.4. Theorem. Let $\mathcal{B}_1$ and $\mathcal{B}_2$ be bases for topologies on X_1 and X_2, respectively. Then

$$\{\pi_1^{-1}(B_1), \pi_2^{-1}(B_2) : B_1 \in \mathcal{B}_1, B_2 \in \mathcal{B}_2\}$$

is a sub-basis for the product topology on $X_1 \times X_2$.

Thus the collection of all strips over basis elements is a sub-basis for the product topology. It is often necessary to have a basis for the product topology (instead of a sub-basis), and we can easily get two such bases.

1.5. Theorem. Let $(X_1, \mathfrak{J}_1)$ and $(X_2, \mathfrak{J}_2)$ be topological spaces. Then
1) $\{U_1 \times U_2 : U_1 \in \mathfrak{J}_1 \text{ and } U_2 \in \mathfrak{J}_2\}$ is a basis for the product topology on $X_1 \times X_2$.
2) If $\mathcal{B}_1$ and $\mathcal{B}_2$ are bases for $\mathfrak{J}_1$ and $\mathfrak{J}_2$ respectively, then $\{B_1 \times B_2 : B_1 \in \mathcal{B}_1 \text{ and } B_2 \in B_2\}$ is a basis for the product topology on $X_1 \times X_2$.
[*Hint for proof:* See Problem 2 in Exercises 1.3.]

Thus the collection of products of open sets is a basis for the product topology, as is the collection of products of basic open sets. It follows immediately, then, that *the product of open sets is open* in the product of two spaces. But it must be emphasized that *an open set in the product of two spaces need not be the product of two open sets* Rather, a set is open in the product topology if it is the *union* of sets which are the product of open sets, and such a union need not itself be the product of open sets.

A similar statement holds for closed sets in the product topology.

1.6. Theorem. Let X_1 and X_2 be topological spaces and let $X_1 \times X_2$ have the product topology. Then
1) If F_1 is closed in X_1 and F_2 is closed in X_2, then $F_1 \times F_2$ is closed in $X_1 \times X_2$.
2) A closed subset of $X_1 \times X_2$ need not be the product of closed sets.

The operation of forming the closure of a set behaves very well with respect to product spaces. Specifically,

1.7. Theorem. Let X_1 and X_2 be topological spaces and let $X_1 \times X_2$ have the product topology. For $A_1 \subset X_1$ and $A_2 \subset X_2$ (with A_1 and A_2 both nonempty), $\bar{A}_1 \times \bar{A}_2 = \overline{A_1 \times A_2}$.

Thus "the product of the closures is the closure of the product." In general, however, if $A \subset X_1 \times X_2$, $\bar{A}$ need not be the product of the closures of two sets.

If $A_1 \subset X_1$ and $A_2 \subset X_2$ then $A_1 \times A_2 \subset X_1 \times X_2$. When X_1 and X_2 are topological spaces, then A_1 and A_2 are also topological spaces with the relative topology inherited from X_1 and X_2. Then $A_1 \times A_2$ can be given the product topology obtained by using the topologies on A_1 and A_2. But $A_1 \times A_2$ can also be given a topology by considering it as a subspace of the product space $X_1 \times X_2$. These two different approaches to giving $A_1 \times A_2$ a topology are actually the same, as the following theorem shows.

1.8. Theorem. Let X_1 and X_2 be topological spaces and let $A_1 \subset X_1$, $A_2 \subset X_2$. The set $A_1 \times A_2$ can be given a topology in two ways:

Method 1) Regard A_1 and A_2 as subspaces of X_1 and X_2, respectively, and give $A_1 \times A_2$ the product topology obtained from the topologies on A_1 and A_2.

Method 2) Give $A_1 \times A_2$ the relative topology inherited from the topology on $X_1 \times X_2$.

Methods 1 and 2 yield the same topology on $A_1 \times A_2$.

Thus "the product of subspaces is a subspace of the product." This result is very useful because it is often the case that one method of looking at the topology on $A_1 \times A_2$ is easier than the other.

Of course not every subspace of $X_1 \times X_2$ is of the form $A_1 \times A_2$ for some $A_1 \subset X_1$ and some $A_2 \subset X_2$. In the general case that a subset S of $X_1 \times X_2$ is not of the form $A_1 \times A_2$, then the only method that can be used to make S into a subspace of $X_1 \times X_2$ is Method 2, giving it the relative topology inherited from the product topology on $X_1 \times X_2$.

Recall that the projection functions on $\mathbf{R} \times \mathbf{R}$ are continuous (Theorem 8.8 in Chapter 4). Now that we have a topology on the product of any two topological spaces, we can prove continuity of the projection functions in general.

1.9. Theorem. Let X_1 and X_2 be topological spaces and let $X_1 \times X_2$ have the product topology. Then the projection functions $\pi_1: X_1 \times X_2 \to X_1$ and $\pi_2: X_1 \times X_2 \to X_2$, defined by $\pi_1(x_1, x_2) = x_1$ and $\pi_2(x_1, x_2) = x_2$, are continuous.

Because they are continuous, we will usually call the projection functions **projection maps.** We can use these projection maps to discuss continuity of functions whose range is a subspace of a product space. Recall that the composition of continuous functions is continuous (Theorem 9.10 in Chapter 4), but that the converse of this statement is false in general. There is one important case, however, when the converse is true. Specifically,

1.10. Theorem. Let X_1, X_2, and Y be topological spaces and let $f : Y \to X_1 \times X_2$, where $X_1 \times X_2$ has the product topology. Then f is continuous if and only if $\pi_1 \circ f$ and $\pi_2 \circ f$ are continuous, where π_1 and π_2 are the projection maps on $X_1 \times X_2$.

Thus a function into a product space is continuous if and only if its composition with both of the projection functions is continuous.

Note that when $f : Y \to X_1 \times X_2$ then $\pi_1 \circ f : Y \to X_1$ and $\pi_2 \circ f : Y \to X_2$. Thus when we have a function into the product $X_1 \times X_2$, we can use the projection functions to get functions into each of X_1 and X_2. The opposite situation (when a function is defined coordinatewise), is also of interest.

1.11. Theorem. Let X_1, X_2, and Y be topological spaces, and let $f_1 : Y \to X_1$, $f_2 : Y \to X_2$. Define $f : Y \to X_1 \times X_2$ by $f(x) = (f_1(x), f_2(x))$. Then f is continuous if and only if both f_1 and f_2 are continuous. The function f is often denoted by $f_1 \times f_2$.

As we observed in our previous discussion of Cartesian products in Chapter 1, X_1 and X_2 are not subsets of $X_1 \times X_2$. However, they are *topological* subsets, in the sense that each is topologically identical to a subspace of $X_1 \times X_2$. Recall that two topological spaces are topologically identical if there is a homeomorphism between them, and that a homeomorphism between two spaces is a 1-1, onto, continuous function whose inverse is also continuous.

1.12. Theorem. Let X_1 and X_2 be topological spaces and let $X_1 \times X_2$ have the product topology. Then X_1 and X_2 are each homeomorphic to a subspace of $X_1 \times X_2$.

[*Hint for proof:* Recall our blanket assumption that X_1 and X_2 are not empty (indeed, the theorem is false if exactly one of them is empty, a fact that you should verify). Let $x_2 \in X_2$ and consider $X_1 \times \{x_2\}$ with the relative topology inherited from $X_1 \times X_2$, a space which looks very much like X_1. Similarly for X_2.]

The spaces $X_1 \times \{x_2\}$ and $\{x_1\} \times X_2$ used in the proof of Theorem 1.12 are called **slices** in $X_1 \times X_2$, parallel to X_1 and X_2, respectively.

1.13. Exercises.

1) If X_1 and X_2 are discrete spaces, then $X_1 \times X_2$ is also discrete.

2) Recall that a function between two topological spaces is *open* (an *open function*) if the image of every open set is open, and a function between two topological spaces is *closed* (a *closed function*) if the image of every closed set is closed.

   Prove that the projection maps are open but not closed.

3) Recall that the interior of a subset A of a topological space X is defined by

$$A° = \cup\{U \subset X : U \text{ is open and } U \subset A\}.$$

Prove that $A° \times B° = (A \times B)°$.

4) Recall that the *derived set* of a subset A of a topological space X is the set A' of all points $x \in X$ such that every neighborhood of x meets A in a point other than x itself.
Is $A' \times B' = (A \times B)'$?

5) Let d be a metric on a set X, so that d is a function from $X \times X$ into **R**. Show that when X has the metric topology induced by d and when $X \times X$ has the product topology, then $d : X \times X \to$ **R** is continuous, when **R** has its usual topology.

6) Show that each factor in $X \times Y$ is a retract of $X \times Y$.

2. FINITE PRODUCTS

Everything that we did for the product of two topological spaces carries over to the product of any finite number of spaces (but not to the product of infinitely many spaces, as we will see in the next section). The key to extending the proof from two spaces to any finite number is usually mathematical induction.

2.1. Definition. Let n be a positive integer greater than 1 and let X_1, $X_2, \cdots, X_n$ be sets. The Cartesian product of these n sets is

$$\prod_{i=1}^{n} X_i = X_1 \times X_2 \times \cdots \times X_n = \{(x_1, x_2, \cdots, x_n) : x_i \in X_i, 1 \leq i \leq n\}^*.$$

Thus the Cartesian product of n sets is simply the set of n-tuples with first coordinate from X_1, second coordinate from X_2, and so on. The sets $X_1, X_2, \cdots, X_n$ are called the **factors** in the product $\prod_{i=1}^{n} X_i$. Two n-tuples are equal if and only if they are equal coordinatewise.

2.2. Definition. Let $n > 1$ and let $X_1, X_2, \cdots, X_n$ be sets. There are n **projection functions** defined on $\prod_{i=1}^{n} X_i$ by

$$\pi_i(x_1, x_2, \cdots, x_n) = x_i, \text{ for } 1 \leq i \leq n.$$

The topology on $\prod_{i=1}^{n} X_i$ is just what we would expect.

2.3. Definition. Let $n > 1$ and let $(X_1, \mathfrak{I}_1), (X_2, \mathfrak{I}_2), \cdots, (X_n, \mathfrak{I}_n)$ be topological spaces. The product topology on $\prod_{i=1}^{n} X_i$ is the topology generated

---

* As with the product of two spaces, the assumption that each $X_i \neq \emptyset$ implies that $\prod_{i=1}^{n} X_i \neq \emptyset$.

by the sub-basis

$$\{\pi_i^{-1}(U_i): U_i \text{ is open in } X_i, 1 \le i \le n\}.$$

As in the case of the product of two spaces, we need not use all of the open sets in the factors to describe a sub-basis for the product topology.

2.4. Theorem. Let $n > 1$ and let $\mathcal{B}_1$, $\mathcal{B}_2$, $\cdots$, $\mathcal{B}_n$ be bases for the topologies on X_1, X_2, $\cdots$, X_n. Then

$$\{\pi_i^{-1}(B_i): B_i \in \mathcal{B}_i, 1 \le i \le n\}$$

is a sub-basis for the product topology on $\prod_{i=1}^n X_i$.

2.5. Exercises.
1) Prove that if $U_i \subset X_i$ for $1 \le i \le n$, then $\pi_i^{-1}(U_i) = X_1 \times X_2 \times \cdots \times X_{i-1} \times U_i \times X_{i+1} \times \cdots \times X_n$.
2) Show that if $U_i \subset X_i$ for $1 \le i \le n$, then $\bigcap_{i=1}^n \pi_i^{-1}(U_i) = U_1 \times U_2 \times \cdots \times U_n = \prod_{i=1}^n U_i$.

The set of "open rectangles" is a basis for the topology on the product of two spaces; the set of "open boxes" is a basis for the topology on the product of finitely many spaces in general, as follows.

2.6. Theorem. Let $n > 1$ and let $(X_1, \mathcal{J}_1)$, $(X_2, \mathcal{J}_2)$, $\cdots$, $(X_n, \mathcal{J}_n)$ be topological spaces. Then
1) $\{\prod_{i=1}^n U_i: U_i \in \mathcal{J}_i\}$ is a basis for the product topology on $\prod_{i=1}^n X_i$.
2) If $\mathcal{B}_i$ is a basis for $\mathcal{J}_i$, $1 \le i \le n$, then $\{\prod_{i=1}^n B_i: B_i \in \mathcal{B}_i\}$ is a basis for the product topology on $\prod_{i=1}^n X_i$.

2.7. Theorem. Let X_1, X_2, $\cdots$, X_n be topological spaces and let $\prod_{i=1}^n X_i$ have the product topology. Then
1) The product of open sets is open.
2) An open set in the product need not be the product of open sets.
3) The product of closed sets is closed.
4) A closed set in the product need not be the product of closed sets.
5) If $A_i \subset X_i$, $1 \le i \le n$, then $\prod_{i=1}^n \bar{A}_i = \overline{\prod_{i=1}^n A_i}$, i.e., the "product of the closures is the closure of the product."

2.8. Theorem. Let X_1, X_2, $\cdots$, X_n be topological spaces and let $A_i \subset X_i$, $1 \le i \le n$. Then the topology on $\prod_{i=1}^n A_i$ as a subspace of the product space $\prod_{i=1}^n X_i$ is the same as the topology on $\prod_{i=1}^n A_i$ when it is viewed as the product of the subspaces A_i with the product topology.

Of course not every subspace of the product space $\prod_{i=1}^n X_i$ is of the form $\prod_{i=1}^n A_i$ for some $A_i \subset X_i$, and in general the topology on a subspace of

$\prod_{i=1}^{n} X_i$ can only be obtained by using the relative topology inherited from $\prod_{i=1}^{n} X_i$. But it is important to know that whenever a subspace of a product space is the product of subspaces, then the topology on the product is the product of the subspace topologies.

2.9. Theorem. Let $n > 1$. The projection functions on a product space with n factors are continuous.

As mentioned before, continuous functions are often called *maps*, and we will usually refer to the projection functions as *projection maps*.

2.10. Theorem. Let $n > 1$ and let X_1, X_2, $\cdots$, X_n be topological spaces. Let Y be a topological space and let $f: Y \to \prod_{i=1}^{n} X_i$. Then f is continuous if and only if for each i, $1 \leq i \leq n$, $\pi_i \circ f$ is continuous.

Thus a function into a product space is continuous if and only if its composition with each of the projection functions is continuous.

2.11. Theorem. Let $n > 1$ and let X_1, X_2, $\cdots$, X_n be topological spaces. Let Y be a topological space and for each i, $1 \leq i \leq n$, let $f_i: Y \to X_i$. Then the function $f: Y \to \prod_{i=1}^{n} X_i$ defined by $f(x) = (f(x_1), f(x_2), \cdots, f(x_n))$ is continuous if and only if each f_i is continuous. The function f is often denoted by $\prod_{i=1}^{n} f_i$.

As with the product of two sets, none of the factors is a subset of the product in any finite product. But each factor is homeomorphic to a subspace of the product space.

2.12. Theorem. Let $n > 1$ and let X_1, X_2, $\cdots$, X_n be topological spaces. Then each X_i is homeomorphic to a subspace of the product space $\prod_{i=1}^{n} X_i$.

2.13. Exercises.

In these exercises, X_1, X_2, $\cdots$, X_n are topological spaces and $\prod_{i=1}^{n} X_i$ has the product topology. See Exercises 1.13 for definitions that are not familiar.

1) If each X_i is a discrete space, then $\prod_{i=1}^{n} X_i$ is also discrete.
2) Is the converse to Problem 1 true?
3) Prove that the projection functions are open but not closed.
4) Denote the interior of a set $A_i \subset X_i$ by A_i°. Then $\prod_{i=1}^{n} A_i^\circ = (\prod_{i=1}^{n} A_i)^\circ$.
5) Denote the derived set of $A_i \subset X_i$ by A_i'. Is $\prod_{i=1}^{n} A_i' = (\prod_{i=1}^{n} A_i)'$?
6) Let S^1 denote the unit circle as a subspace of the Euclidean plane with its usual topology, and let T denote the (hollow) torus as a subspace of Euclidean 3-space with the product topology that it gets by virtue of being $\mathbf{R} \times \mathbf{R} \times \mathbf{R}$ (where each copy of $\mathbf{R}$ has its

usual topology). (The torus is the surface of a doughnut, or an inner tube.) Let $S^1 \times S^1$ have the product topology, and show that $S^1 \times S^1$ and T are homeomorphic.

7) Show that each factor is a retract of $\prod_{i=1}^{n} f_i$.

3. INFINITE PRODUCTS

Before defining the product of infinitely many factors, let us review the case of just two factors. We said that

$$X_1 \times X_2 = \{(x_1, x_2) : x_1 \in X_1 \text{ and } x_2 \in X_2\}.$$

In other words, $X_1 \times X_2$ is the set of all ordered pairs with first coordinate from X_1 and second coordinate from X_2. Another way to look at such an ordered pair is as a function (of all things). Indeed, (x_1, x_2) can be thought of as the function from $\{1, 2\}$ to $X_1 \cup X_2$ that sends 1 to x_1 and 2 to x_2. In symbols, $(x_1, x_2) : \{1, 2\} \to X_1 \cup X_2$ by $(x_1, x_2)(1) = x_1$ and $(x_1, x_2)(2) = x_2$. Thus we could have defined $X_1 \times X_2$ as follows:

$$X_1 \times X_2 = \{x : \{1, 2\} \to X_1 \cup X_2 : x(1) \in X_1, x(2) \in X_2\},$$

the set of all functions $x : \{1, 2\} \to X_1 \cup X_2$ such that $x(1) \in X_1$ and $x(2) \in X_2$. If we call $\{1, 2\}$ the **index set** for $X_1 \times X_2$, we can say that $X_1 \times X_2$ is the set of all functions x from the index set to the *union* of the factors such that $x(i) \in X_i$ for each i in the index set. This is the approach that we will use to define Cartesian products in general.

3.1. Definition. Let Λ be any non-empty set such that for each $\lambda \in \Lambda$ there is a non-empty set X_λ. (In other words $\{X_\lambda : \lambda \in \Lambda\}$ is a collection of non-empty sets indexed by the set Λ.) The **Cartesian product** of $\{X_\lambda : \lambda \in \Lambda\}$ is defined to be

$$\prod_{\lambda \in \Lambda} X_\lambda = \{x : \Lambda \to \textstyle\bigcup_{\lambda \in \Lambda} X_\lambda : x(\lambda) \in X_\lambda \text{ for each } \lambda \in \Lambda\},$$

the set of all functions x from the index set Λ to the union of the factors $(\bigcup_{\lambda \in \Lambda} X_\lambda)$, such that for each $\lambda \in \Lambda$, $x(\lambda) \in X_\lambda$.*

* Unlike the case of finitely many factors, the fact that the product of infinitely many factors is non-empty when each of the factors is non-empty is not a trivial matter in general. In the general case of uncountably many factors, it is a consequence of the *axiom of choice* which says that given any non-empty collection of non-empty sets, there exists a set consisting of exactly one element from each set in the collection. The axiom of choice is not as "obviously true" as the other axioms of set theory (such as "the intersection of two sets is a set") which we have used without comment. But if we are going to deal with infinite sets in general (which are not so obvious themselves), we are going to need something like the axiom of choice. There is a very interesting discussion of the axiom of choice and some equivalences of it in Wilder [18]. (As a matter of fact, that the general Cartesian product of non-empty sets is non-empty is not just a consequence of the axiom of choice but is equivalent to it.)

On the other hand, the fact that the product is empty whenever at least one of the factors is empty is obvious and does not require the axiom of choice.

The value $x(\lambda)$ of the function (point in the product) x is usually denoted by x_λ and is called the λ-th **coordinate** of the point x. The point x is usually written $x = (x_\lambda)$ or sometimes $x = (x_\lambda)_{\lambda \in \Lambda}$. When Λ is countably infinite, we will often write $\prod_{\lambda \in \Lambda} X_\lambda$ as $\prod_{i=1}^{\infty} X_i$, and will write $(x_\lambda)_{\lambda \in \Lambda}$ as $(x_1, x_2, \cdots)$.

Fortunately, it is not always necessary to think of a point in the product as a function. Rather, in most cases, we can think of the product space (rather imprecisely) as the set of all "Λ-tuples," $(x_\lambda)_{\lambda \in \Lambda}$, the set of all points whose λ-th coordinate is an element of X_λ.

The projection functions on a general Cartesian product are defined as follows.

3.2. Definition. Let $\{X_\lambda : \lambda \in \Lambda\}$ be a non-empty collection of non-empty sets. For each $\lambda_0 \in \Lambda$ there is a **projection function** $\pi_{\lambda_0} : \prod_{\lambda \in \Lambda} X_\lambda \to X_{\lambda_0}$, defined by

$$\pi_{\lambda_0}(x) = \pi_{\lambda_0}((x_\lambda)) = x_{\lambda_0}.$$

Thus the λ_0-th projection function sends the product onto its λ_0-th factor.

We define a topology on the general Cartesian product by copying what we did in the finite case.

3.3. Definition. Let $\{(X_\lambda, \Im_\lambda) : \lambda \in \Lambda\}$ be a non-empty collection of non-empty topological spaces. The **product topology** on $\prod_{\lambda \in \Lambda} X_\lambda$ is the topology generated by the sub-basis

$$\{\pi_\lambda^{-1}(U_\lambda) : U_\lambda \in \Im_\lambda\}$$

Thus, as with finite products, the topology on the general Cartesian product is generated by the sub-basis of all "strips" over open sets in each factor. Of course we need not use *all* open sets to describe a sub-basis for the product topology, as the following theorem shows.

3.4. Theorem. Let $\{(X_\lambda, \Im_\lambda) : \lambda \in \Lambda\}$ be a non-empty collection of non-empty topological spaces. If $\mathfrak{B}_\lambda$ is a basis for $\Im_\lambda$ for each $\lambda \in \Lambda$, then the collection

$$\{\pi_\lambda^{-1}(B_\lambda) : B_\lambda \in \mathfrak{B}_\lambda\}$$

is a sub-basis for the product topology on $\prod_{\lambda \in \Lambda} X_\lambda$.

Unlike the case of finite products, the product of non-trivial open sets in an infinite product is *not* open! To prove it, remember that a set is open if and only if each point in it is contained in a sub-basic open set which is totally contained in the set. Read the definition of sub-basis very carefully.

3.5. Theorem. Let $\{X_\lambda : \lambda \in \Lambda\}$ be a non-empty collection of non-empty topological spaces. If $|\Lambda| \geq \aleph_0$ and if, for each $\lambda \in \Lambda$, U_λ is open in X_λ, then $\prod_{\lambda \in \Lambda} U_\lambda$ is open in the product topology on $\prod_{\lambda \in \Lambda} X_\lambda$ if and only if $U_\lambda = X_\lambda$ for all but finitely many $\lambda \in \Lambda$.

3.6. Corollary. Let $\{X_\lambda : \lambda \in \Lambda\}$ be a non-empty collection of non-empty topological spaces. If $|\Lambda| \geq \aleph_0$ and, for each $\lambda \in \Lambda$, U_λ is open in X_λ with $U_\lambda \neq \emptyset$ and $U_\lambda \neq X_\lambda$, then $\prod_{\lambda \in \Lambda} U_\lambda$ is not open in the product space $\prod_{\lambda \in \Lambda} X_\lambda$.

When an open set U_λ is such that $U_\lambda \neq \emptyset$ and $U_\lambda \neq X_\lambda$, U_λ is called a *non-trivial* open set. Thus the collection of products of open sets is *not* a basis for the product topology on the product of infinitely many factors (as it *is* on the product of finitely many factors) because, in the case of infinitely many factors, the product of non-trivial open sets is not even open. This raises a legitimate question: why define the product topology in such a way that the product of open sets is not open? It *could* be defined so that the product of open sets is open, but it is not. Why not?

One reason is that we want to copy the sub-basis of open "strips" that we used in the case of finitely many factors, and in the resulting topology on infinitely many factors, the product of non-trivial open sets is simply not open—an unfortunate consequence. Why not then copy instead the idea that the collection of all products of open sets is a *basis* for the product topology (as it is in the case of finitely many factors), because then the product of open sets would be open automatically (any basic open set is certainly an open set). The topology on a product obtained by using a basis of products of open sets has been studied (it is called the "box topology" or the "box product") but it is not the same as the product topology on a product with infinitely many factors.

A good reason for defining the product topology as we did has to do with the amount of "control" over things that it is reasonable to demand. For example, consider the sequence $\{1/n : n \in \mathbf{Z}^+\} \subset \mathbf{R}$. This sequence converges to 0 because given any neighborhood of 0, no matter how small, the sequence is ultimately in the neighborhood. Thus determining whether a sequence converges or not involves a degree of control over the sequence in the sense that we check to see if it gets close and stays close to a particular point (0 in the case of $\{1/n : n \in \mathbf{Z}^+\}$). Another way to state the definition of convergence is to say that a sequence converges to a point if it is ultimately in the interesection of any *finite* collection of neighborhoods of that point; but if we try to control it too much by saying that it must ultimately be in the intersection of *all* neighborhoods of the point, then we will destroy the intuitive idea of convergence entirely. Indeed, the intersection of all neighborhoods of 0 in $\mathbf{R}$ is the singleton set $\{0\}$, and $\{1/n : n \in \mathbf{Z}^+\}$ is not only not *ultimately* in this intersection, it is *never* in this intersection. Thus trying to control things too much can destroy the way they ought to be.

This example can be applied to product spaces as follows.

3.7. Exercise.

Let $\{X_n : n \in \mathbf{Z}^+\}$ be a countable collection with each $X_n = \mathbf{R}$ with the usual topology. Consider the sequence $\{(1/n, 1/n, \cdots) : n \in \mathbf{Z}^+\} \subset \prod_{i=1}^{\infty} X_i$. Certainly this sequence should converge to the point $(0, 0, \cdots) \in \prod_{i=1}^{\infty} X_i$. But if we try to "over-control" the product space by requiring that the product of open sets be open, it will not. For each $n \in \mathbf{Z}^+$, put $U_n = (-1/n, 1/n)$.

1) Show that each U_n is open in $\mathbf{R}$, but the sequence $\{(1/n, 1/n, \cdots) : n \in \mathbf{Z}^+\}$ is never in $\prod_{i=1}^{\infty} U_i$, so if this set is open in the product, $\{(1/n, 1/n, \cdots) : n \in \mathbf{Z}^+\}$ will *not* converge to $(0, 0, \cdots)$, which is not reasonable.

2) Show that $\{(1/n, 1/n, \cdots) : n \in \mathbf{Z}^+\}$ *does* converge to $(0, 0, \cdots)$ when $\prod_{i=1}^{\infty} X_i$ has the product topology.

Thus the definition of the product topology seems to be more reasonable than the box topology, even though the product of infinitely many non-trivial open sets is not open in the product topology.

Since the product of open sets in general is not open, the collection of all products of open sets is not a basis for the topology on the product of infinitely many factors, as it is on the product of finitely many factors. Theorem 3.9 below gives a basis for the topology on an infinite product. First, though, we need a simple lemma.

3.8. Lemma. Let $\{X_\lambda : \lambda \in \Lambda\}$ be a non-empty collection of non-empty sets with $|\Lambda| \geq \aleph_0$. Then for $\lambda_1, \lambda_2, \cdots, \lambda_n \in \Lambda$, $n \in \mathbf{Z}^+$, if $U_{\lambda_i} \subset X_{\lambda_i}$, $\bigcap_{i=1}^{n} \pi_{\lambda_i}^{-1}(U_{\lambda_i}) = \pi_{\lambda \in \Lambda} A_\lambda$, where $A_\lambda = \begin{cases} U_{\lambda_i} \text{ if } \lambda = \lambda_i \\ X_\lambda \text{ if } \lambda \neq \lambda_i \end{cases}$, $1 \leq i \leq n$.

3.9. Theorem. Let $\{X_\lambda : \lambda \in \Lambda\}$ be a non-empty collection of non-empty topological spaces, with $|\Lambda| \geq \aleph_0$.

1) If for each $\lambda \in \Lambda$, U_λ is open in X_λ, then the collection

$$\{\textstyle\prod_{\lambda \in \Lambda} U_\lambda : \text{all but finitely many } U_\lambda = X_\lambda\}$$

is a basis for the product topology on $\prod_{\lambda \in \Lambda} X_\lambda$.

2) If for each $\lambda \in \Lambda$, $\mathcal{B}_\lambda$ is a basis for the topology on X_λ, then if $X_\lambda \in \mathcal{B}_\lambda$ for each $\lambda \in \Lambda$, the collection

$$\{\textstyle\prod_{\lambda \in \Lambda} B_\lambda : B_\lambda \in \mathcal{B}_\lambda \text{ and all but finitely many } B_\lambda = X_\lambda\}$$

is a basis for the product topology on $\prod_{\lambda \in \Lambda} X_\lambda$.

Thus the collection of products of non-empty open sets, only finitely many of which are non-trivial, is a basis for the product topology on the product of infinitely many factors. It is often useful to write such a basis element as $\bigcap_{i=1}^{n} \pi_{\lambda_i}^{-1}(U_{\lambda_i})$.

The fact that the product of non-empty open sets is open if and only if only finitely many of them are non-trivial is often expressed by saying that *an open set in a product space restricts only finitely many coordinates.* In terms of projection functions, this becomes

3.10. Theorem. Let $\{X_\lambda : \lambda \in \Lambda\}$ be a non-empty collection of non-empty topological spaces, and let $\prod_{\lambda \in \Lambda} X_\lambda$ have the product topology. If U is *any* open subset of $\prod_{\lambda \in \Lambda} X_\lambda$, then $\pi_\lambda(U) = X_\lambda$ for all but finitely many $\lambda \in \Lambda$.

Closed sets behave "better" than open sets in the product topology:

3.11. Theorem. Let $\{X_\lambda : \lambda \in \Lambda\}$ be a non-empty collection of non-empty topological spaces, and for each $\lambda \in \Lambda$ let F_λ be a closed subset of X_λ. Then $\prod_{\lambda \in \Lambda} F_\lambda$ is closed in $\prod_{\lambda \in \Lambda} X_\lambda$ with the product topology.

Of course not every closed subset of a product is a product of closed sets, but it is still important to know that the product of closed sets is a closed set. Similarly, "the product of the closures is the closure of the product." More precisely,

3.12. Theorem. Let $\{(X_\lambda, \mathfrak{I}_\lambda) : \lambda \in \Lambda\}$ be a non-empty collection of non-empty topological spaces, and for each $\lambda \in \Lambda$, let $A_\lambda \subset X_\lambda$. Then $\prod_{\lambda \in \Lambda} \bar{A}_\lambda = \overline{\prod_{\lambda \in \Lambda} A_\lambda}$.

As we saw with finite products, if a subspace of a product is the product of subspaces, then it can be given a topology in two ways, and the two topologies are identical. The same thing happens with infinite products.

3.13. Theorem. Let $\{X_\lambda : \lambda \in \Lambda\}$ be a non-empty collection of non-empty topological spaces. For each $\lambda \in \Lambda$, let A_λ be a subspace of X_λ. Then the product topology on $\prod_{\lambda \in \Lambda} A_\lambda$ is the same as the relative topology on $\prod_{\lambda \in \Lambda} A_\lambda$ inherited from the product topology on $\prod_{\lambda \in \Lambda} X_\lambda$.

Thus, as with finite products, "the product of subspaces is a subspace of the product." Of course not every subspace of a product space is the product of subspaces, and in general, a subspace can only be given the relative topology in one way.

The projection functions on an infinite product are continuous, and we can use them as before to discuss the continuity of a function whose range is a subspace of a product space.

3.14. Theorem. Let $\{X_\lambda : \lambda \in \Lambda\}$ be a non-empty collection of non-empty topological spaces. For each $\lambda_0 \in \Lambda$, the projection function $\pi_{\lambda_0} : \prod_{\lambda \in \Lambda} X_\lambda \to X_{\lambda_0}$ defined by $\pi_{\lambda_0}(x_\lambda) = x_{\lambda_0}$, is continuous.

3.15. Theorem. Let $\{X_\lambda : \lambda \in \Lambda\}$ be a non-empty collection of non-empty topological spaces, and let Y be a topological space. Let $f : Y \to \prod_{\lambda \in \Lambda} X_\lambda$

where $\prod_{\lambda \in \Lambda} X_\lambda$ has the product topology. Then f is continuous if and only if for each $\lambda \in \Lambda$, $\pi_\lambda \circ f$ is continuous.

3.16. Theorem. Let $\{X_\lambda : \lambda \in \Lambda\}$ be a non-empty collection of non-empty topological spaces, let $\prod_{\lambda \in \Lambda} X_\lambda$ have the product topology, and let Y be a topological space. For each $\lambda \in \Lambda$, let $f_\lambda : Y \to X_\lambda$. Then the function $f : Y \to \prod_{\lambda \in \Lambda} X_\lambda$ defined by $f(x) = (f_\lambda(x))_{\lambda \in \Lambda}$ is continuous if and only if each f_λ is continuous. The function f is often denoted by $\prod_{\lambda \in \Lambda} f_\lambda$.

As with finite products, none of the factors of an infinite product is a subset of the product. But each factor is homeomorphic to a subspace of the product space, as the following theorem shows.

3.17. Theorem. Let $\{X_\lambda : \lambda \in \Lambda\}$ be a non-empty collection of non-empty topological spaces. For each $\lambda \in \Lambda$, X_λ is homeomorphic to a subspace of $\prod_{\lambda \in \Lambda} X_\lambda$.

3.18. Exercises.

In these exercises, $\{X_\lambda : \lambda \in \Lambda\}$ is a non-empty collection of non-empty topological spaces, and $\prod_{\lambda \in \Lambda} X_\lambda$ has the product topology. See Exercises 1.13 for definitions of terms that are not familiar.

1) If $|\Lambda| \geq \aleph_0$, $\prod_{\lambda \in \Lambda} X_\lambda$ is *never* a discrete space no matter what the X_λ's are.
2) The projection maps on $\prod_{\lambda \in \Lambda} X_\lambda$ are open but not closed.
3) Is the product of the interiors equal to the interior of the product?
4) If $A_\lambda \subset X_\lambda$ for each $\lambda \in \Lambda$, is $\prod_{\lambda \in \Lambda} A_\lambda' = (\prod_{\lambda \in \Lambda} A_\lambda)'$?
5) Show that each factor is a retract of $\prod_{\lambda \in \Lambda} X_\lambda$.

chapter six

Connectivity

Connectivity is an important concept in analysis and topology. Intuitively, a set is connected if it is all in one piece, and you have already encountered the concept of connectivity in the intermediate value theorem in calculus. This theorem says essentially that if $f : I \to \mathbf{R}$ is continuous (where I is an interval), and if c is any number between any two points $f(a)$ and $f(b)$ in the range of f, then there is a point x between a and b in the domain such that $f(x) = c$. In other words, when the domain of a continuous function is all in one piece, then the range is also all in one piece: the continuous image of a connected set is connected.

In this chapter we will investigate connectivity in general and will show, among other things, that a theorem like the intermediate value theorem is true in general: the continuous image of a connected set is always connected.

1. DEFINITION OF CONNECTIVITY AND SOME USEFUL CONSEQUENCES

We have said that the intuitive idea of connectivity is that a subset of a topological space is connected if it is all in one piece. However, our intuition is based on years of working with the real line with its usual topology, and intuition can be misleading in a more general setting. For example, the set $(0, 1) \cup (1, 2)$ is not all in one piece and is not a connected subset of $\mathbf{R}$ with its usual topology. But in the real line with the indiscrete topology, the set $(0, 1) \cup (1, 2)$ is connected. On the other hand, $(0, 1)$ is connected in $\mathbf{R}$ with the usual topology but is not connected in $\mathbf{R}$ with the discrete topology. (Of course we cannot prove any of these statements until we get a definition of connectivity.)

Basically, a set is connected if it cannot be split into more than one piece by disjoint open sets. This is a negative idea: we say what it means for a set to be connected by specifying what cannot happen. Our formal definition is also given negatively. Furthermore, we will be interested in the connectivity of subsets of topological spaces as well as in connectivity of the entire space. It turns out, though, that the definition of connectivity is more obviously what it should be if we give it for the entire space rather than for a subset of a space; connectivity of a subset is then determined by viewing the subset as a topological space itself with the relative topology.

1.1. Definition.
1) A topological space X is **connected** if it cannot be written as the union of two non-empty disjoint open sets.
2) A subset A of a topological space X is connected if it is connected as a topological space when given the relative topology inherited from the topology on X.

This definition says exactly what connectivity ought to be. A space would have to be in at least two pieces in order to be equal to the union of two *non-empty disjoint* open sets, because each of the open sets would have to contain part of the set. Key words in the definition are *non-empty* and *disjoint*; if either of them is omitted, the definition would no longer say what we would want connectivity to mean.

To prove that a space is connected according to the definition involves showing that it *cannot* be written as the union of two non-empty disjoint open sets. The only way to do this is to assume that it *can* be so written and then show that this assumption leads to a contradiction. On the other hand, to prove that a space is not connected, all you need to do is to exhibit two non-empty disjoint open sets whose union is the space in question.

To prove some of our previous statements,

1.2. Exercises.
1) *Every* subset (including the space itself) of an indiscrete space is connected.
2) A subset of a discrete space is connected if and only if it is a singleton set. In particular, a discrete space with more than one point is not connected.
3) The set $(0, 1) \cup (1, 2)$ is not a connected subset of **R** with its usual topology.
4) The empty set is a connected subset of every topological space.

We also claimed earlier that the interval $(0, 1)$ is a connected subset of **R** with its usual topology. The proof of this is more difficult than the other statements and we postpone it until we have more information about connectivity than just the definition.

As with continuity, there are several equivalent formulations of the definition of connectivity. These are useful to have because one way of looking at connectivity is often easier to use than another. A word about proving that a series of statements is an equivalent series of statements is in order. In general, two statements are *equivalent* if they say exactly the same thing. More precisely, two statements are equivalent if each implies the other. Thus to prove that two statements are equivalent, *assume* that one of them holds and use it to deduce the other; then reverse the process. To prove that more than two statements are equivalent, we could look at them pairwise and prove that each pair is equivalent, but this is usually more work than is necessary. For example, suppose that we have three statements A, B, and C that we want to prove equivalent. We could show that A and B are equivalent and that B and C are equivalent. Then we could deduce easily that since A implies B and B implies C (because A is now equivalent to B and B equivalent to C), then A implies C; similarly, we get C implies A. The two implications between A and C then allow us to say that A and C are equivalent. This process involves four real proofs (A implies B, B implies A, B implies C, C implies B), which is less than the six proofs that would be required to show that all possible pairs of A, B, and C are equivalent. But we can usually reduce the work even more by setting up a chain of implications: A implies B, B implies C, C implies A. This involves only three proofs and establishes the equivalence of the three statements A, B, and C because it shows that each implies the other. One more thing: the order in which the statements are given is not necessarily the easiest order in which to prove any implications between them. The best idea is to find the easy proofs first (if there are any), prove them, and then fit the others in with as little extra work as possible. Sometimes two simple proofs are easier than one hard one, especially if you don't see how to do the hard one right away, so some of the statements may fit into the chain as "branches" attached to only one link. For example, to prove five statements A, B, C, D, E equivalent, we might end up with something like

$$
\begin{array}{ccc}
& B \Rightarrow A & \\
\nearrow & & \Downarrow \\
D \Leftarrow & C \Leftarrow E &
\end{array}
\qquad \text{or, maybe,} \qquad
\begin{array}{ccc}
A \Rightarrow & C & \Leftrightarrow E \\
\Uparrow & & \Downarrow \\
B \Leftarrow & D &
\end{array}
$$

($\Rightarrow$ means *implies:* $\Leftrightarrow$ means *is equivalent to*. Note that "$A \Leftrightarrow B$" is the same as "A if and only if B.")

There are many other possibilities of course. The only thing that matters is to end up with each statement implying the other.

Finally, we should note that to prove that two statements A and B are equivalent, we do not care in the least whether or not they are true statements. To prove equivalence, assume A holds (whether it does or not is really irrelevant), and then deduce that B follows from A. Then assume B and deduce A.

1.3. Theorem. Let X be a topological space. The following statements are equivalent.
   1) X cannot be written as the union of two non-empty disjoint open sets (i.e., X is connected according to our definition).
   2) X cannot be written as the union of two non-empty disjoint closed sets.
   3) X does not contain a non-empty proper subset which is both open and closed in X.
   4) X cannot be written as $H \cup K$ where both H and K are non-empty, and both $\bar{H} \cap K = \emptyset$ and $H \cap \bar{K} = \emptyset$.

Since most of our discussion of connectivity will concern subsets of a space rather than the whole space, a formulation of the definition of connectivity of a subset directly (rather than saying that a subset is connected if it is connected as a subspace) will be useful. We give such a definition in the following theorem. Its proof is a direct application of the definitons of connectivity of a topological space and of the relative topology.

1.4. Theorem. Let X be a topological space. A non-empty subset A of X is connected if and only if there do not exist open subsets U and V of X such that $A \cap U \neq \emptyset$, $A \cap V \neq \emptyset$, $A \cap U \cap V = \emptyset$ and $A \subset U \cup V$.

Notice that in Theorem 1.4 we do not require that U and V be disjoint, but only that they not overlap in A. What happens to them outside of A has no bearing on whether or not A is connected.

Theorem 1.4 gives us one way to show that a subset A of a topological space is connected: assume that A is contained in $U \cup V$, where U and V are open subsets of the space, and then show that either $A \cap U = \emptyset$ or $A \cap V = \emptyset$. As with connectivity of a space itself, there are several equivalent ways to determine connectivity of a subset of a space. Those corresponding to the statements in Theorem 1.3 are given below.

1.5. Theorem. Let X be a topological space and let A be a non-empty subset of X. The following statements are equivalent.
   1) A is connected, i.e., if $A \subset U \cup V$ where U and V are non-empty disjoint open subsets of X, then either $A \cap U = \emptyset$ or $A \cap V = \emptyset$.
   2) If $A \subset F_1 \cap F_2$ where F_1 and F_2 are non-empty disjoint closed subsets of X, then either $A \cap F_1 = \emptyset$ or $A \cap F_2 = \emptyset$.
   3) If Y is a non-empty proper subset of X which is both open and closed in X, then $Y \cap A = \emptyset$, or $Y = A$.

4) A cannot be written as $H \cup K$, where $H \cap A \neq \emptyset$, $K \cap A \neq \emptyset$, and $\bar{H} \cap K = H \cap \bar{K} = \emptyset$.

To prove that a subset of **R** with its usual topology is connected if and only if it is an interval, we need some information about the real line which may already be familiar.

1.6. Definitions.

1) A subset A of **R** is **bounded above** if there exists a positive integer n such that $A \subset (-\infty, n]$; A is **bounded below** if there exists a positive integer n such that $A \subset [-n, \infty)$. A subset A of **R** is **bounded** if it is both bounded above and bounded below.

2) Let A be a non-empty subset of **R** which is bounded above. The **least upper bound** of A is the number $r \in \mathbf{R}$ such that $r \geq a$ for all $a \in A$, and if $t \in \mathbf{R}$ is such that $t \geq a$ for all $a \in A$, then $t \geq r$. The least upper bound of A is denoted by **supA,** and is called the **supremum** of A.

3) Let A be a non-empty subset of **R** which is bounded below. The **greatest lower bound** of A is the number $r \in \mathbf{R}$ such that $r \leq a$ for all $a \in A$ and if $t \in \mathbf{R}$ is such that $t \leq a$ for all $a \in A$, then $t \leq r$. The greatest lower bound of A is called the **infimum** of A and is denoted by **infA.**

Thus a subset A of **R** is bounded above if it does not "run off to $+\infty$," is bounded below if it does not "run off to $-\infty$," and is bounded if it does not "run off to infinity in either direction." It is a fundamental property of **R** that if A is a non-empty subset of **R**, then sup$A$ exists if and only if A is bounded above, and inf$A$ exists if and only if A is bounded below. (This property is called the **completeness** property of **R**. We will return to it later.) When they exist, sup$A$ is the smallest real number which is greater than or equal to every element of A, and inf$A$ is the largest real number which is less than or equal to every element of A.

1.7. Exercise

1) What are sup$A$ and inf$A$ when A is
   a) $(0, 1)$.
   b) $[0, 1)$.
   c) $(0, 1]$.
   d) $[0, 1]$.
   e) $\{r \in \mathbf{Q} : r < \sqrt{2}\}$.
   f) $\{1/n : n \in \mathbf{Z}^+\}$.

2) (Problems 2 and 3 require a knowledge of ordinal numbers. In particular, See sections 5 and 6 of Chapter 2.)
   We can talk about least upper bounds and greatest lower bounds

in some sets other than $\mathbf{R}$. For example, sup$A$ and inf$A$ can be defined in the obvious way (if they exist) when $A \subset [0, \Omega)$. Do they always exist? How about if A is countable?

  3) What are sup$A$ and inf$A$ if $A \subset [0, \Omega)$ is
    a) $[0, \omega)$.
    b) $\{2n : n \in \mathbf{Z}^+\}$.
    c) $\{2^n : n \in \mathbf{Z}^+\}$. The least upper bound of this set might be called 2^ω; is $2^{|\omega|} = |\,2^\omega\,|$?

The following theorem relates least upper bounds and greatest lower bounds to the usual topology on $\mathbf{R}$.

1.8. Theorem. Let $\mathbf{R}$ have the usual topology and let A be a non-empty subset of $\mathbf{R}$.
  1) If A is bounded above (i.e., if sup$A$ exists), then sup$A \in \bar{A}$.
  2) If A is bounded below (i.e., if inf$A$ exists), then inf$A \in \bar{A}$.

1.9. Corollary. Let $\mathbf{R}$ have the usual topology and let F be a non-empty closed subset of $\mathbf{R}$. Then
  1) If F is bounded above (i.e., if sup$F$ exists), then sup$F \in F$.
  2) If F is bounded below (i.e., if inf$F$ exists), then inf$F \in F$.

1.10. Exercises.
In these exercises, $\mathbf{R}$ has the usual topology.
  1) Is the following statement true or false: If $F \subset \mathbf{R}$, $F \neq \emptyset$, and F is bounded, then F is closed if and only if both sup$F \in F$ and inf$F \in F$.
  2) We have not defined sup$\emptyset$ or inf$\emptyset$. Extend the definition of least upper bound and greatest lower bound of a set to apply to the empty set. [*Hint*: Neither sup$\emptyset$ nor inf$\emptyset$ can be a real number.]
  3) Give an example to show that each of the following statements is false.
    a) If $F \subset \mathbf{R}$, $F \neq \emptyset$ and F is closed, then sup$F \in F$.
    b) If $F \subset \mathbf{R}$, $F \neq \emptyset$ and F is closed, then inf$F \in F$.

Problem 3 above is important. Before we can talk about the least upper bound or greatest lower bound of a subset of $\mathbf{R}$, we must be sure that they *exist*, and this means that we must be sure that the subset is properly bounded (bounded above for the sup, bounded below for the inf). You can run into trouble rapidly if you are not careful about this, and it is very little trouble to make sure that the set in question is bounded.

We can finally prove that a subset of $\mathbf{R}$ (with the usual topology) is connected if and only if it is an interval. Remember that $I \subset \mathbf{R}$ is an interval if and only if for any two points a, $b \in I$, *all* points between a and b also belong to I. (Thus, in particular, the empty set is an interval.)

1.11. Theorem. A subset of **R** (with the usual topology) is connected if and only if it is an interval.

[*Hint for proof:* For the "only if" part, see the remark above about intervals. For the "if" part, suppose that $I \subset$ **R** is an interval that is not connected. Then I can be written as the union of two disjoint, non-empty, closed sets. If these two closed sets are intervals, the rest of the proof is easy. However, they need not be intervals: If they are not, then they "interlock." Consider the sup of a piece of one of them and the inf of the piece of the other above it. Use Corollary 1.9.]

In particular, Theorem 1.11 shows that **R** itself is connected when it has its usual topology, because $\mathbf{R} = (-\infty, \infty)$, so is an interval.

1.12. Corollary. The only subsets of **R** with its usual topology that are both open and closed are the empty set and **R** itself.

2. PRESERVATION OF CONNECTIVITY BY CERTAIN OPERATIONS AND FUNCTIONS

It is easy to see that the union of two disjoint connected sets need not be connected. But if they overlap, then the union is connected. Before we can prove it, we need a simple modification of Theorem 1.4.

2.1. Lemma. If A is a non-empty connected subset of a topological space X and $A \subset U \cup V$ where U and V are disjoint open subsets of X, then either $A \subset U$ or $A \subset V$.

2.2. Theorem. Let A and B be connected subsets of a topological space X. If $A \cap B \neq \emptyset$ then $A \cup B$ is connected.

Theorem 2.2 can be extended to arbitrary unions as follows.

2.3. Theorem. Let $\{A_\lambda : \lambda \in \Lambda\}$ be a collection of connected subsets of a topological space X. If $\bigcap_{\lambda \in \Lambda} A_\lambda \neq \emptyset$ then $\bigcup_{\lambda \in \Lambda} A_\lambda$ is connected.

2.4. Exercises.
  1) The intersection of two connected sets need not be connected. [*Hint:* Look in the plane.]
  2) If A and B are both connected, $A - B$ need not be connected.

If a set is connected then it is all in one piece. It should seem reasonable that if we adjoin all points that are close to a connected set to the set, then the resulting set is still connected. In other words, the closure of a connected set should be connected. In fact, we can prove the following.

2.5. Theorem. Let X be a topological space. If $A, B \subset X$ such that $A \subset B \subset \bar{A}$, then if A is connected, B is also connected.

[*Hint for proof:* Suppose that B is not connected. Use Lemma 2.1 and the definition of closure.]

2.6. Corollary. The closure of a connected set is connected.

As mentioned earlier, connectivity is preserved by continuous functions.

2.7. Theorem. Let X and Y be topological spaces and let $f: X \rightarrow Y$ be continuous. If X is connected, then $f(X)$ is also connected.

Theorem 2.7 is one of the most important results that we have obtained so far, for several reasons. One is that it will enable us to derive other important facts, such as the fact that the product of (non-empty) topological spaces is connected if and only if each factor is connected. Another is that it will give us a formulation of the definition of connectivity in terms of continuity, something that will often be useful. Yet another reason has to do with homeomorphism. Since a homeomorphism is first of all a continuous function, then when two spaces are homeomorphic, the homeomorphism between them is a continuous function (which is also 1-1, onto, and has a continuous inverse). In particular, though, it is continuous: so, according to Theorem 2.7, a connected space *cannot* be homeomorphic to a space that is not connected. Thus a connected space and a non-connected space are never topologically the same. We should note that if $f: X \rightarrow Y$ is continuous and onto, then if X is connected, Y is also connected, even if f is not a homeomorphism. However, it is possible for $f: X \rightarrow Y$ to be continuous and onto with Y connected and X *not* connected. If we want to guarantee that when f is a function from X onto Y, then X and Y are *both* connected or *both* not connected, then f must be a homeomorphism from X onto Y.

We should also note that Theorem 2.7 shows that a disconnected space cannot be a retract of a connected space. Thus, for example, $\{0, 1\}$ is not a retract of $\mathbf{R}$, or of $[0, 1]$.

For the characterization of connectivity in terms of continuity, recall that a discrete space with more than one point is not connected (Problem 2 of Exercises 1.2). Then

2.8. Theorem. A topological space X is connected if and only if there does not exist a continuous function from X onto $\{0, 1\}$ with the discrete topology.

To prove that a product of non-empty spaces is connected if and only if each of the factors is connected, we have to draw several facts together. Recall that the projection functions are continuous and that the continuous image of a connected space is connected. Recall also that each factor is homeomorphic to a subset of the product and that the union of connected spaces which all have a point in common is connected. Putting these facts together in the right way will enable you to prove the following theorem. (Drawing a picture might help, too.)

2.9. Theorem. Let X and Y be non-empty topological spaces and let $X \times Y$ have the product topology. Then $X \times Y$ is connected if and only if both X and Y are connected.

By Theorem 2.9, the Euclidean plane with its usual topology is connected because $\mathbf{R} = (-\infty, \infty)$ is connected (Theorem 1.11), and when the plane is viewed as $\mathbf{R} \times \mathbf{R}$, the product topology on the plane is the same as the usual topology on the plane. This is an example of how it is often easier to do something in general than it is to do it directly for a specific case. When you think about it, proving directly (without using product spaces) that the plane with its usual topology is connected might be complicated. (Proving directly that $\mathbf{R}$ is connected was not easy; the plane could be even worse.) But the general fact that the product of two connected spaces is also connected is not very hard to prove, and the special case that the plane is connected follows immediately (once we know that $\mathbf{R}$ is connected.) As a corollary to the fact that the plane is connected, we have the following.

2.10. Theorem. The only subsets of the Euclidean plane (with its usual topology) that are both open and closed are the empty set and the plane itself.

By using mathematical induction and Theorem 2.9, you can show that the product of any finite number of non-empty connected spaces is connected if and only if each factor is connected.

2.11. Theorem. Let $n \in \mathbf{Z}^{+}$ and let $\{X_i : 1 \leq i \leq n\}$ be a finite collection of non-empty topological spaces. Let $\prod_{i=1}^{n} X_i$ have the product topology. Then $\prod_{i=1}^{n} X_i$ is connected if and only if each X_i is connected, $1 \leq i \leq n$.

The rest of this section (except for Theorem 2.18) requires a knowledge of infinite products.

To prove that an arbitrary product of non-empty topological spaces is connected if and only if each factor is connected is more difficult than the finite case, at least in one direction. As is often the case with an "if and only if" theorem (which of course always requires two proofs, one for the "if" part and one for the "only if" part), one proof is easy and one is quite difficult. The easy one here is the "only if" part, which we prove now.

2.12. Theorem. Let $\{X_\lambda : \lambda \in \Lambda\}$ be a non-empty collection of non-empty topological spaces and let $\prod_{\lambda \in \Lambda} X_\lambda$ have the product topology. If $\prod_{\lambda \in \Lambda} X_\lambda$ is connected, then each X_λ is also connected.

For the "if" part, we need a new concept. Recall that in **R** with its usual topology, the closure of the set of rational numbers is **R** itself. In other words, every real number is close to a rational number. We generalize this as follows.

2.13. Definition. Let X be a topological space. A subset A of X is **dense** in X if $\bar{A} = X$.

The following characterization of a dense subset of a topological space follows directly from the definition of a dense subset and the definition of closure.

2.14. Theorem. Let X be a topological space. A subset A of X is dense in X if and only if every non-empty open subset of X meets A (i.e., for every non-empty open subset U of X, $U \cap A \neq \emptyset$).

In product spaces,

2.15. Theorem. Let $\{X_\lambda : \lambda \in \Lambda\}$ be a non-empty collection of non-empty topological spaces, let $\prod_{\lambda \in \Lambda} X_\lambda$ have the product topology and let x be any point of $\prod_{\lambda \in \Lambda} X_\lambda$. Then the set of all points of $\prod_{\lambda \in \Lambda} X_\lambda$ that differ from x in only finitely many coordinates is dense in $\prod_{\lambda \in \Lambda} X_\lambda$.

[*Hint for proof:* Any open set contains a basic open set, so if every basic open set meets a given set, so does every open set. Use Theorem 2.14.]

If we can show that the set of all points that differ from a fixed $x \in \prod_{\lambda \in \Lambda} X_\lambda$ in only finitely many coordinates is a connected set, then since this set is dense in $\prod_{\lambda \in \Lambda} X_\lambda$, it will follow from Corollary 2.6 that $\prod_{\lambda \in \Lambda} X_\lambda$ is connected. To do this, we need the following generalization of Theorem 1.12 in Chapter 5.

2.16. Lemma. Let $\{X_\lambda : \lambda \in \Lambda\}$ be a non-empty collection of non-empty topological spaces, let $x \in \prod_{\lambda \in \Lambda} X_\lambda$ and let n be a positive integer. For $\lambda_1, \lambda_2, \cdots, \lambda_n$ put

$$A_\lambda = \begin{cases} X_\lambda & \text{if } \lambda = \lambda_i, \ 1 \le i \le n \\ \{x_\lambda\} & \text{if } \lambda \neq \lambda_i, \ 1 \le i \le n \end{cases}.$$

Then $\prod_{\lambda \in \Lambda} A_\lambda$ with the product topology is homeomorphic to $\prod_{i=1}^{n} X_{\lambda_i}$ with the product topology.

Now we can prove that a product of non-empty connected spaces is connected.

2.17. Theorem. Let $\{X_\lambda : \lambda \in \Lambda\}$ be a non-empty collection of non-empty topological spaces. If each X_λ is connected, then $\prod_{\lambda \in \Lambda} X_\lambda$ with the product topology is also connected.

Outline of Proof: Let $x \in \prod_{\lambda \in \Lambda} X_\lambda$ and for $n \in \mathbf{Z}^+$, choose $\lambda_1, \lambda_2, \cdots, \lambda_n \in \Lambda$. Define $\{A_\lambda : \lambda \in \Lambda\}$ as in Lemma 2.16 and deduce that $S(x : \lambda_1, \lambda_2, \cdots, \lambda_n) = \prod_{\lambda \in \Lambda} A_\lambda$ is connected. Show that the union of all possible sets $S(x : \lambda_1, \lambda_2, \cdots, \lambda_n)$ with x fixed and $\lambda_1, \lambda_2, \cdots \lambda_n$ ranging over *all* finite subsets of Λ is connected and is dense in $\prod_{\lambda \in \Lambda} X$. Conclude that $\prod_{\lambda \in \Lambda} X_\lambda$ is connected.

Combining Theorems 2.12 and 2.17, we have the fact that the product of non-empty topological spaces is connected if and only if each factor is connected.

As another application of Theorem 2.7, we can prove the intermediate value theorem.

2.18. Theorem. Let $\mathbf{R}$ have its usual topology, let $I \subset \mathbf{R}$ be an interval and let $f : I \to \mathbf{R}$ be a continuous function. For any two numbers a and b in I and any number c between $f(a)$ and $f(b)$ in the range of f, there is a number $x \in I$ such that $f(x) = c$.

3. COMPONENTS AND LOCAL CONNECTIVITY

As we have seen, $(0, 1) \cup (1, 2)$ is not a connected subset of $\mathbf{R}$ with its usual topology. However, both $(0, 1)$ and $(1, 2)$ are connected, and, furthermore, for each point in $(0, 1)$, $(0, 1)$ is the largest connected subset of $(0, 1) \cup (1, 2)$ containing the point. Similarly, $(1, 2)$ is the largest connected subset of $(0, 1) \cup (1, 2)$ containing each of its points. In general,

3.1. Definition. Let X be a topological space and let $x \in X$. The largest connected subset of X containing x is called the **component** of X that contains x, and is denoted by C_x.

Thus for $x \in X$, C_x is a connected subset of X containing x, such that if C is a connected subset of X which also contains x, then $C \subset C_x$. When dealing with components it is often easier to use the following theorem instead of the definition.

3.2. Theorem. Let X be a topological space and let $x \in X$. Then

$$C_x = \bigcup \{K \subset X : K \text{ is connected and } x \in K\}.$$

Thus the component of X containing a given point is the union of all connected subsets of X that contain the point.

When $Y \subset X$ is a subspace of X, the component of Y containing a given point $y \in Y$ is the largest connected subset *of* Y containing y, and is the union of all connected subsets *of* Y that contain y.

3.3. Exercises.

1) Let X be a discrete space, $x \in X$. What is C_x? Is it closed? Is it open?

2) Let X be a discrete space, $Y \subset X$ and $y \in Y$. What is the component of y in Y? Is it closed in Y? Is it open in Y? Is it closed in X? Is it open in X?

3) Let $x \in \mathbf{Q}$ where $\mathbf{Q}$ is the set of rational numbers with the relative topology inherited from the usual topology on $\mathbf{R}$. What is C_x? Is it closed? Is it open? Is $\mathbf{Q}$ a discrete space?

4) Let $x \in X$ where X is a connected topological space. What is C_x? If $y \in X$, $y \neq x$, what is C_y?

5) Let $\mathbf{R}$ have its usual topology, and let A be a subspace of $\mathbf{R}$ containing 0 (i.e., A has the relative topology inherited from the usual topology on $\mathbf{R}$ and $0 \in A$). What is the component of A containing 0 if A is

a) $\mathbf{R}$.

b) $\mathbf{Z}$.

c) $\mathbf{Q}$.

d) $\{1/n : n \in \mathbf{Z}^+\} \cup \{0\}$.

e) $\mathbf{R} - \{1/n : n \in \mathbf{Z}^+\}$.

6) Let E^2 be the plane with its usual topology and let A be a subspace of E^2 containing $(0, 0)$. What is the component of A containing $(0, 0)$ if A is

a) E^2.

b) $\mathbf{R}$.

c) $\{(x, y) : x, y \in \mathbf{Q}\}$.

d) $E^2 - \{(x, y) : x, y \in \mathbf{Q}, x \neq 0, y \neq 0\}$.

As Problem 4 above shows, a connected topological space contains only one component, namely itself. A space whose components are all singletons is called **totally disconnected.** Thus a space is totally disconnected if and only if $C_x = \{x\}$ for all $x \in X$. Problem 1 above shows that every discrete space is totally disconnected, but, as Problem 3 shows, not every totally disconnected space is discrete.

Using components allows us to tell how disconnected a space is by breaking it up into connected pieces—its components. The components of a space *partition* the space in the following sense.

3.4. Definition. Let X be a set. A collection of non-empty subsets $\{A_\lambda : \lambda \in \Lambda\}$ **partitions** X if

1) $\bigcup_{\lambda \in \Lambda} A_\lambda = X$, and

2) $A_\lambda \cap A_\mu = \emptyset$ whenever $\lambda, \mu \in \Lambda$ with $\lambda \neq \mu$.

We call the collection $\{A_\lambda : \lambda \in \Lambda\}$ a **partition** of the set X.

3.5. Theorem. The collection of components of a topological space is a partition of the space.

Thus the components of a topological space decompose the space into pairwise disjoint connected subsets. It is easy to see from the definition of a component and Corollary 2.6 that each component of a space is closed. For later reference, we state this simple fact as a theorem.

3.6. Theorem. Each component of a topological space X is a closed set.

At first thought, it might seem that the components of a topological space should be open sets as well. Indeed, if a space has only finitely many components, then each of them is open (Why?), in addition to being closed. However, if a space has infinitely many components, then they need not be open (see Problem 3 in Exercises 3.3).

The components of a topological space decompose it into pairwise disjoint, closed, connected subsets, and tell us how disconnected it is. Since connectivity is preserved by a homeomorphism, it should not be too surprising that the "amount of disconnectivity" is also preserved by a homeomorphism. In other words, two homeomorphic spaces are either both connected or are disconnected to the same extent. The following theorem shows that a homeomorphism between two spaces must "match up" the components of the two spaces.

3.7. Theorem. Let X and Y be topological spaces and let $h: X \to Y$ be a homeomorphism of X onto Y (i.e., X and Y are homeomorphic). Then for each $x \in X$, $h(C_x) = C_{h(x)}$ and for each $y \in Y$, $h^{-1}(C_y) = C_{h^{-1}(y)}$.

Thus a homeomorphism between two spaces X and Y induces a 1-1 correspondence between the components of X and the components of Y, and in fact, corresponding components under this 1-1 correspondence are homeomorphic. Hence, as mentioned earlier, homeomorphic topological spaces are either both connected or have exactly the same "amount of disconnectivity." It is easy to see, though, that having the same "amount of disconnectivity" is not enough to guarantee that two topological spaces are the same (Exercise 3.8 (1) below). What may be surprising is that we can have a 1-1 correspondence between the components of two spaces (so that they have the same "amount of disconnectivity") and we can even have corresponding components homeomorphic, while the spaces themselves are not homeomorphic (Exercise 3.8 (2) below).

3.8. Exercises.

1) The same "amount of disconnectivity" does not guarantee homeomorphism: let $X = (0, 1) \cup (1, 2)$ with the relative topology inherited from the usual topology on **R** and let $Y = \{a, b\}$ with the

discrete topology. Show that there is a 1-1 correspondence between the components of X and the components of Y, but X and Y are not homeomorphic.

2) A 1-1 correspondence between components (same "amount of disconnectivity") with corresponding components homeomorphic does not guarantee homeomorphic spaces: Recall the 1-1 correspondence between the positive integers and the positive rational numbers (see Theorem 1.7 in Chapter 2). Give both sets the relative topology inherited from the usual topology on **R**, and show that there is a 1-1 correspondence between the components of these two spaces, with corresponding components homeomorphic, but the two spaces are not homeomorphic.

3) (This example is due to C. Kuratowski, and is taken from Dugundji ([4], p. 112, Ex. 6).) Let X and Y be subspaces of **R** (with its usual topology) defined by

$$X = (0, 1) \cup \{2\} \cup (3, 4) \cup \{5\} \cup (6, 7) \cup \{8\} \cup \cdots$$

$$Y = (0, 1] \cup (3, 4) \cup \{5\} \cup (6, 7) \cup \{8\} \cup \cdots$$

a) Show that X and Y are not homeomorphic. [*Hint:* Components must be homeomorphic if the spaces are.]

b) Show that $f: X \to Y$ defined by

$$f(x) = \begin{cases} x & \text{if } x \neq 2 \\ 1 & \text{if } x = 2 \end{cases}$$

is continuous, 1-1, and onto.

c) Show that $g: Y \to X$ defined by

$$g(y) = \begin{cases} x/2 & \text{if } x \in (0, 1] \\ (x - 2)/2 & \text{if } x \in (3, 4) \\ x - 3 & \text{otherwise} \end{cases}$$

is continuous, 1-1, and onto.

Recall the Schröder-Bernstein theorem (Theorem 1.5 in Chapter 2). This example shows that a similar theorem fails for homeomorphism: if there is a continuous 1-1 onto function from X to Y and there is also a continuous 1-1 onto function from Y to X, there need *not* be a homeomorphism between X and Y.

We have seen that the components of a topological space are closed sets but are not necessarily open. When certain of them are open, we can derive some important consequences.

3.9. Definition. A topological space X is **locally connected** if the components of each open subset of X are open.

This definition is not very descriptive of the "local-ness" of local connectivity. The following theorem makes it more apparent.

3.10. Theorem. A topological space X is locally connected if and only if for each point $x \in X$, every open neighborhood of x contains a *connected* open neighborhood of x.

[*Hint for proof:* Recall that a set is open if and only if for each point in it, it contains an open set that contains the point.]

According to Theorem 3.10, a space is locally connected if and only if it has a local basis consisting of *connected* open sets at each of its points.

To get an idea of what to look for when considering local connectivity (or the lack of it), consider the following subspace of the plane, the "broom."

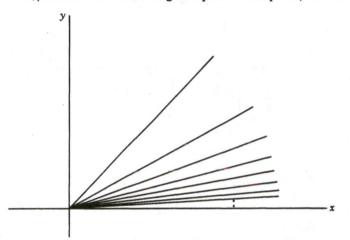

The broom space is a subspace of the plane consisting of a sequence of lines of length 1 emanating from the origin with slopes $\frac{1}{2}$, $\frac{1}{3}$, $\frac{1}{4}$, $\cdots$, together with $[0, 1]$ on the x-axis.

We claim that the broom space is not locally connected. To convince yourself of this, take any point $(x, 0)$ in $(0, 1]$, and observe that a basic open neighborhood of it will have to look like

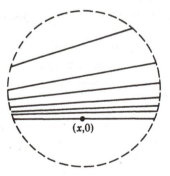

so is a set consisting of disconnected line segments, and cannot contain a connected neighborhood of the point.

It may be surprising, but there is absolutely no relation between connectivity and local connectivity, as we show in the following exercises.

3.11. Exercises.

1) Does the broom space above fail to be locally connected at every one of its points?
2) Is the broom space above connected?
3) Consider the closed "topologist's sine curve"

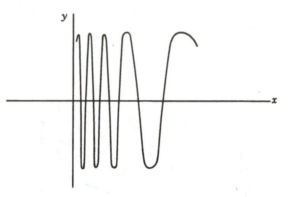

consisting of the graph of $y = \sin(1/x)$, $0 < x \le 1$ (called the topologist's sine curve) together with its "limit line," the segment from -1 to 1 on the x-axis. The topology is the relative topology inherited from the plane.

    a) Show that this space is *not* locally connected.
    b) Show that this space *is* connected. [*Hint:* See Theorem 2.7 and Corollary 2.6.]

4) Give an example of a locally connected space which is not connected. [*Hint:* This is easy.]
5) Show that a discrete space is always locally connected. Is a discrete space ever connected?
6) Show that an indiscrete space is always locally connected. Is an indiscrete space ever connected?
7) According to the definition, the components of each open subset of a locally connected space are open. Show that the components of a non-open subset of a locally connected space need not be open. [*Hint:* Look in **R**.]

Unlike connectivity, local connectivity is not necessarily preserved by continuous functions, as you can show in the following exercise.

3.12. Exercise.

Let $X = \mathbf{Z}^+$ and $Y = \{1/n : n \in \mathbf{Z}^+\} \cup \{0\}$, both with the relative topology inherited from the usual topology on $\mathbf{R}$. Define $f : X \to Y$ by $f(1) = 0$ and $f(n) = 1/(n - 1)$ for $n > 1$. Show that f is continuous, that X is locally connected and Y is not. Thus the continuous image of a locally connected space need not be locally connected.

Even though local connectivity is not necessarily preserved by continuous functions, it is preserved by a homeomorphism: local connectivity is a topological property.

3.13. Theorem. If two topological spaces are homeomorphic, then one of them is locally connected if and only if the other one is also locally connected.

Local connectivity is also preserved by finite products. In the case of two factors,

3.14. Theorem. Let X and Y be two non-empty topological spaces and let $X \times Y$ have the product topology. Then $X \times Y$ is locally connected if and only if both X and Y are locally connected.

[*Hint for proof:* Remember that the product of two open sets is open, the product of two connected sets is connected, and the projection maps are open maps.]

By induction, we can extend the result of Theorem 3.14 to the product of any finite number of spaces.

3.15. Theorem. Let $n > 1$ and let $\{X_1, X_2, \cdots, X_n\}$ be a finite collection of non-empty topological spaces. Let $\prod_{i=1}^{n} X_i$ have the product topology. Then $\prod_{i=1}^{n} X_i$ is locally connected if and only if each X_i is locally connected.

The rest of this section requires a knowledge of infinite products.

The key to the proof that the product of finitely many locally connected spaces is locally connected is the fact that in the product of finitely many spaces, the product of open sets is open. In the case of infinitely many factors, the product of open sets need not be open, so the method of proof used in the finite case will fail in the infinite case. As a matter of fact, the result is not even true in the case of infinitely many factors: the arbitrary product of locally connected spaces need not be locally connected, as the following exercise shows.

3.16. Exercises.

Let $X = \{0, 1\}$ with the discrete topology and let $Y = \prod_{i=1}^{\infty} X_i$, where each $X_i = X$. Let Y have the product topology.

1) Show that X is locally connected (so each factor of Y is locally connected), but
2) Show that Y is not locally connected. [*Hint:* Y is not discrete (Why not?), so some points of Y are not open sets. (In fact, *none* of the points of Y are open (Why not?).) Show that Y is totally disconnected, i.e., that the components of Y are singleton sets. Conclude that Y is not locally connected.]

The problem with the space Y in Exercises 3.16 is that too many of its factors are not connected, and since an open set in a product can restrict only finitely many coordinates, Y does not contain any connected open sets. To get the product of locally connected spaces locally connected, Y shows us that we are going to have to put more restrictions on most of the factors, as in the following theorem.*

3.17. Theorem. Let $\{X_\lambda : \lambda \in \Lambda\}$ be a non-empty collection of non-empty topological spaces and let $\prod_{\lambda \in \Lambda} X_\lambda$ have the product topology. Then $\prod_{\lambda \in \Lambda} X_\lambda$ is locally connected if and only if each X_λ is locally connected and all but finitely many of the X_λ are also connected.

4. OTHER KINDS OF CONNECTIVITY

We have said that a subset of a topological space is connected if it is all in one piece, and have made this idea precise by saying that such a subset is connected if it is not contained in the union of two disjoint open sets, each of which meets the subset in question. There is another way to make precise the idea of a set's being all in one piece, a way that may seem at first to be more intuitively "right" (but which, incidentally, is not equivalent to connectivity as we have defined it). We might say that a set is all in one piece if any two points in the set can be joined together by an unbroken (but probably very crooked) "line." To make this more precise, we need to know in general what such a "line" is.

Throughout this section, let I denote the unit interval $[0, 1]$ with its usual topology, the relative topology inherited from the usual topology on **R**.

4.1. Definition. A **path** in a topological space X is a continuous function $p : I \to X$. The **initial point** of the path p is the point $p(0)$, and the **terminal point** is $p(1)$; we say that the path $p : I \to X$ runs from $p(0)$ to $p(1)$.

For $A \subset X$, a path in A is a continuous function $p : I \to A$ where A has the relative topology inherited from the topology on X.

---

* Strictly speaking, the additional restriction is not necessary for the "only if" part of the theorem, but stating this part as a theorem by itself without the additional restriction would be talking about nothing when Λ is infinite, because there is no locally connected product with infinitely many non-connected factors.

The image of a path p in X is the subset $\{p(t):t \in I\}$ of X, and is called a **curve** in X.

Thus a path in a subspace A of a space X is a continuous function from I into A. It is important to remember that the path is the function itself, rather than the image of I under this function. The reason for this distinction is that we want to be able to talk about the *direction* of a path and say that it goes from one point to another, and we want to be able to do this uniquely and unambiguously. Thus, for example, the path in the plane going from $(1, 1)$ to $(2, 2)$ given by $p(t) = (t + 1, t + 1)$ for $t \in I$ is not the same as the path in the plane going from $(2, 2)$ to $(1, 1)$ given by $q(t) = (2 - t, 2 - t)$ for $t \in I$, even though they both represent the same curve in the plane. (You should draw a picture of this curve in the plane and make sure that you see that both paths give rise to the same curve.) In fact, we might call the path q the inverse of the path p, since going from $(1, 1)$ to $(2, 2)$ by p and then going back by q gets us right back where we started from, along the same route in both directions. In general,

4.2. Definition. The **inverse** of a path $p:I \to X$ is the path $p^{-1}I: \to X$ defined by $p^{-1}(t) = p(1 - t)$ for $t \in I$. (Note that this is different from our previous use of the symbol p^{-1}.)

Thus a path $p:I \to X$ and its inverse $p^{-1}:I \to X$ give the same image of I in X, but the initial point of p^{-1} is the terminal point of p, and the terminal point of p^{-1} is the initial point of p.

For two points x and y in a subset A of a space X, we say that x and y can be **joined by a path** in A if there exists a path (continuous function) $p:I \to A$ such that $p(0) = x$ and $p(1) = y$ (where A has the relative topology inherited from the topology on X).

We can now state precisely the idea of a kind of connectivity (which is not equivalent to connectivity as we know it) involving joining points together by unbroken "lines."

4.3. Definition. A subset A of a topological space X is **path connected** if any two points in A can be joined by a path in A. In particular, a space X is path connected if any two points of X can be joined by a path in X.*

Note the requirement that a subspace A of a topological space X is path connected if any two points of A can be joined by a path *in A*. For example, the subspace $(0, 1) \cup (1, 2)$ of **R** is *not* path connected: the points $\frac{1}{2}$ and $\frac{3}{2}$, for example, cannot be joined by a path in $(0, 1) \cup (1, 2)$. However,

---

* Some authors use the term "arc-wise connected" to mean what we mean by "path connected," while others reserve "arc-wise connected" for a situation when all paths are actually arcs: an *arc* is a homeomorphism defined on [0, 1], while a path need only be a continuous function on [0, 1].

these two points *can* be joined by a path in **R**, but $(0, 1) \cup (1, 2)$ is still not path connected.

4.4. Exercises.

1) A subset of a discrete space is path connected if and only if it is a singleton set. In particular, a discrete space with more than one point is not path connected.

2) Every subset of an indiscrete space is path connected. In particular, every indiscrete space is path connected.

3) A subset of **R** with its usual topology is path connected if and only if it is an interval.

As the above exercises show, path connectivity is the same as connectivity in discrete spaces, indiscrete spaces, and in **R** with its usual topology. However, path connectivity is, in general, a stronger property of a space than connectivity. In other words, every path connected space is connected, but there exist connected spaces which are not path connected, as we will show after we get a useful lemma.

4.5. Lemma. A subset A of a topological space X is path connected if and only if given any point $x_0 \in A$, any other point of A can be joined to x_0 by a path in A.

[*Hint for proof:* First of all, be sure that you see how this lemma differs from the definition of path connectivity.

The "only if" part is trivial. For the "if" part, let x and y be any two points of A. By hypothesis, each can be joined to x_0 by a path in A, say p joins x to x_0 and q joins x_0 to y. Going by p from x to x_0 and then by q from x_0 to y would get you from x to y, but in the process you would have to use the points of I twice. This can be remedied by shrinking the domains of p and q and defining $P:I \to A$ as follows:

$$P(t) = \begin{cases} p(2t) & \text{if } 0 \le t \le \tfrac{1}{2} \\ q(2t - 1) & \text{if } \tfrac{1}{2} \le t \le 1. \end{cases}$$

Show that P is a path in A from x to y and conclude that A is path connected.]

The trick of shrinking the domains of paths and then combining them to get a new path (as in the proof of Lemma 4.5) is a handy one to be aware of. In fact, we often think of t as time, and say that two paths such that the terminal point of one is the initial point of the other can be combined to form a third path by traversing each of the two original paths in half of the normal time. We will use this idea again later.

Now we can prove that every path connected space is also connected.

4.6. Theorem. If a topological space X is path connected, then it is connected.

[*Hint for proof*: Choose $x_0 \in X$. Use Lemma 4.5 and Theorem 2.3.]

An immediate (and very important) consequence of Theorem 4.6 is that if a space is *not* connected, then it is also *not* path connected. However, a space can be connected even though it is not path connected. In other words, the converse to Theorem 4.6 is false, as you can show in the following exercise.

4.7. Exercise.

Recall the closed topologist's sine curve, consisting of the graph of $y = \sin(1/x)$, $0 < x \leq 1$, together with its "limit line," the line from -1 to 1 on the y-axis. The topology is the relative topology inherited from the usual topology on the plane. We will call this space S. (We saw this space before in Exercise 3.11 (3).)

1) Prove again that S is connected.
2) Prove that S is not path connected. [*Hint*: If S is path connected, then any two points of S can be joined by a path. In particular, if S is path connected, there is a path p joining the point $(1/\pi, 0)$ to the point $(0, 0)$.

   Use the fact that $\pi_1 \circ p$ is continuous to show that $\pi_2 \circ p$ is not continuous, which is impossible if p is continuous. (To do this, exploit the fact that the $\sin(1/x)$ part of S is connected (Why is it?) and attains the values 1 and -1 infinitely often in any neighborhood of 0.]

The example in Exercise 4.7 shows one instance of why we want connectivity to be as we defined it, rather than taking "connected" to mean "path connected." If connected is to mean "all in one piece" (and, after all, even "path connected" is an attempt to make the idea of "all in one piece" precise), then the closed topologist's sine curve certainly ought to be connected, because there is no way to make it into two really separate pieces without destroying it topologically. But it is not path connected, so if by "connected" we mean "path connected," the closed topologist's sine curve would not be connected.

Using composition of functions (and, in particular, the fact that the composition of continuous functions is continuous), we can prove that, like connectivity, path connectivity is preserved by continuous functions.

4.8. Theorem. Let X and Y be topological spaces and let $f: X \rightarrow Y$ be continuous. Then if X is path connected, $f(X)$ is also path connected.

In particular, Theorem 4.8 shows that when X and Y are homeomorphic, then X is path connected if and only if Y is path connected.

Combining Theorem 4.8 with Exercise 4.7 and some other relevant facts, you can show the following important difference between connectivity and path connectivity.

4.9. Exercise.

Give an example of a path connected space whose closure is not path connected.

Analogous to the components of a space, we have the following.

4.10. Definition. Let X be a topological space and let $x \in X$. The **path component** of X containing x is the set of all points of X that can be joined to x by a path in X.

The path component of $A \subset X$ containing $x \in A$ is the set of all points of A that can be joined to x by a path in A.

Use the method of proof in Lemma 4.5 to prove the following "maximality" property of path components.

4.11. Theorem. Let X be a topological space, $x \in X$, and let P_x denote the path component of X that contains x. If y is any point of P_x and $z \in X$ can be joined to y by a path in X, then $z \in P_x$.

As with components,

4.12. Theorem. Let X be a topological space.
1) For $x \in X$, the path component of X containing x is the largest path connected subset of X that contains x (largest in the sense of containment).
2) The path components of a space partition it into disjoint non-empty path connected subsets.

4.13. Exercises.
1) What are the path components of the closed topologist's sine curve of Exercise 4.7? What are the components of this space? Are the path components open? Are they closed?
2) What are the path components of a discrete space? What are the components of a discrete space? Are the path components open? Are they closed?
3) What are the path components of a connected subset of **R** with its usual topology?
4) Show that a space is path connected if and only if it has exactly one path component, namely itself. Can a subspace of a path connected space have more than one path component? In other words, is a subspace of a path connected space necessarily path connected?
5) Is the broom space of Exercises 3.11 path connected? How about the broom space with the point $(0, 0)$ deleted? What are its components? What are its path components? Are they open? Are they closed?

As Exercise 4.13 (1) above shows, *unlike* the components of a space, the path components of a space need not be closed, and, *like* the components of a space, the path components of a space need not be open.

We know that when the components of certain subspaces of a space are open (namely the components of each of its open sets), then the space is locally connected, so a connected space is also locally connected if the components of each of its open sets is open. It turns out that a similar condition on the path components of a connected space will ensure that it is also path connected. Specifically,

4.14. Theorem. A connected space is path connected if and only if all of its path components are open.

[*Hint for proof:* First of all, note that the condition is only that the path components of the space itself be open, and not that the path components of every open set (which includes the space itself) be open, so it is similar to but not quite the same kind of condition that guarantees that a connected set is also locally connected. (And of course the condition here is on the path components rather than the components.)

For the proof, the "only if" part is trivial. For the "if" part, note that if each path component is open then each path component is also closed.]

The fact that the conditions that a connected space must satisfy in order to be locally connected are somewhat similar to those that guarantee that a connected space is path connected does *not* mean that there is any relation between local connectivity and path connectivity, as you can show in the following exercises. (Actually, we have already seen examples of what is required in these exercises.)

4.15. Exercises.
1) Give an example of a path connected (and therefore connected) space which is not locally connected.
2) Give an example of a locally connected space which is not path connected.
3) Can there be a connected, locally connected space which is not path connected?

We have observed that a subspace of **R** (with its usual topology) is connected if and only if it is an interval, and a subspace of **R** (with its usual topology) is path connected if and only if it is an interval. Thus connectivity and path connectivity are equivalent in **R** with its usual topology. In the plane with its usual topology, however, we have seen that connectivity and path connectivity are not equivalent. It turns out, though, that an *open* subset of the plane is connected if and only if it is path connected, a fact that we can establish once we show that the plane has a path connected basis of open sets.

4.16. Theorem. Let X and Y be non-empty topological spaces and let $X \times Y$ have the product topology. Then $X \times Y$ is path connected if and only if both X and Y are path connected.

[*Hint for proof:* For (x_1, y_1) and $(x_2, y_2) \in X \times Y$, show that there is a path from (x_1, y_1) to (x_2, y_1) "parallel to X," and there is a path from $(x_., y_1)$ to (x_2, y_2) "parallel to Y." Combine these two paths (as in Lemma 4.5) to get a path from (x_1, y_1) to (x_2, y_2). The converse is trivial.]

Combining Theorem 4.16 with Theorem 4.14, you can show the following.

4.17. Theorem. An open subset of the plane is connected if and only if it is path connected.

4.18. Exercise.

(This exercise requires a knowledge of infinite products.) We have seen that the arbitrary product of connected spaces is connected, but the arbitrary product of locally connected spaces need not be locally connected. Is the arbitrary product of path connected spaces necessarily path connected?

Both connectivity and path connectivity are ways to determine whether or not a space is all in one piece, and path connectivity is a stronger idea than connectivity in the sense that every path connected space is connected but not every connected space is path connected.

If we know that a space is path connected, then there is no question as to its being all in one piece. But there is another question which is related in some way to its connectivity, namely the question of how many "holes" it has in it. For example, consider the subspaces (A) and (B) of the plane that look like

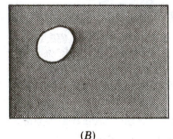

(A)   (B)

Both (A) and (B) are path connected, but (B) has a hole in it, and it turns out that (A) and (B) are *not* topologically the same—they are not homeomorphic. The problem is to be able to recognize the hole in space (B) in some precise way so that we can investigate the consequences of having a hole in a path connected space, and, in particular, see why a path connected space with a hole in it must be topologically different from a path connected space without any holes.

One way to approach this problem is to look at the complements of these two subspaces of the plane. Indeed, $E^2 - (A)$ is connected and $E^2 - (B)$ is not connected ($E^2 - (B)$ is in two pieces: the piece inside the hole and the piece outside the rectangular edge of (B)), so these two complements are topologically different.

As we will see, saying that a space has no holes if its complement is connected is a good idea, but it is not quite good enough. Before examining this further, let us look briefly at another way to decide if a path connected space has any holes, a way that has a great deal of intuitive appeal and that has been the subject of much work in a branch of topology called *algebraic* topology. This is to approach the problem in terms of closed paths (called loops: a loop is a path whose initial point and terminal point are the same). For example, for any point $x_0 \in (A)$, any loop starting and ending at x_0 can be shrunk down to the point x_0 without tearing the loop and without leaving the space (A), as follows (we show several possible stages in the shrinking of the loop to the point):

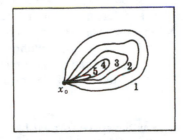

But in space (B), a loop starting and ending at a point $y_0 \in B$ can be shrunk down to the point y_0 without tearing the loop and without leaving the space (B) *if and only if the loop does not enclose the hole* in (B):

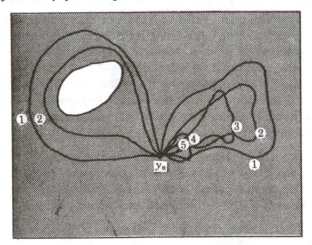

Thus a path connected subspace of the plane with no holes should be one in which given any point in the space, any loop starting and ending at that point can be shrunk to the point without tearing the loop and without leaving the space.

We will explore this fascinating idea of shrinking loops further in the chapter on homotopy. For now let us restrict ourselves to the plane with its usual topology and investigate holes in a path connected subspace of the plane from the point of view of the complement of the space. Path connected subspaces of the plane without any holes are important in complex analysis, where the idea is usually restricted to *open* subsets of the plane. We will observe that restriction here. An open path connected subset of the plane is called a **region** in the plane. (Note that, according to Theorem 4.17, a region in the plane can be described as an open connected subset of the plane.)

As mentioned earlier, saying that a region in the plane has no holes if its complement is connected is not good enough. For example, the region in the plane consisting of the part of the plane outside of a circle has a connected complement, but also has a hole in it (the inside of the circle is a hole in the space). Noting that this space is unbounded and that the "complement is connected" idea of a space without holes seems to work for bounded spaces, we might be tempted to have two definitions of a space without holes, one for bounded subsets of the plane and one for unbounded subsets. However, there is an easier and much more elegant way to deal with this problem.

We can regard the plane as the space formed by removing a single point from a (hollow) sphere and then flattening out the sphere minus this point (you do this every time you look at a (flat) map of the (round) world). Starting with the plane, we can add in a point (called the **point at infinity**) to the plane to retrieve the sphere, and all we have to know in order to do this is where to put this point. Since a point in a topological space is located relative to the other points in the space by its neighborhood basis, all we have to do to position the point at infinity is to specify what neighborhoods of it look alike, and we can do this by copying the way things look on the sphere. When a point is removed from the sphere and the resulting "punctured sphere" is flattened out into the plane, the *complement* of a basic open neighborhood of the missing point becomes a closed and bounded subset of the plane. Thus,

4.19. Definition. The **extended plane**, E^{2*}, is the space $E^2 \cup \{p\}$ where $p \in E^2$, with topology defined as follows: neighborhoods of points in E^2 are the same as in the usual topology on E^2, and $U \subset E^{2*}$ is an open neighborhood of p if and only if $E^{2*} - U$ is closed and bounded in E^2.

Now we can define precisely what it means for a region in the plane to have no holes.

4.20. Definition. A region in the plane is **simply connected** if its complement with respect to the extended plane is connected.

As mentioned earlier, simple connectivity is important in complex analysis, and Definition 4.20 is the way that it is usually defined there.

4.21. Exercises.

Which of the following subsets of the plane is simply connected?

1) The plane itself.
2) A singleton subset of the plane.
3) The "punctured plane," the plane minus one point.
4) The unit square minus two points.
5) The region outside the unit square.

chapter seven

Compactness

Like connectivity, compactness is one of the most important concepts in topology and analysis. One reason for its importance is that a continuous function whose domain is a compact set behaves very well. Indeed, you have already encountered compactness in calculus in the theorem that says that a continuous real-valued function on a closed and bounded interval attains both its maximum and minimum values; since all of the theory of the Riemann integral rests on this theorem, it is one of the most important theorems underlying the calculus. It relates to compactness because a closed and bounded interval is an example of a *compact* subset of **R**.

In this chapter, we will investigate compactness in general. At first you may find the study of compactness more difficult than connectivity, but after you get used to the somewhat strange definition, you should find it a fascinating subject.

1. THE DEFINITION OF COMPACTNESS

The concept of the connectivity of a set arose from an attempt to make precise the idea of a set's being all in one piece, and the resulting definition of connectivity is such that sets that are intuitively all in one piece (such as intervals on the line, for example) turn out to be connected according to the definition. The word "compact" is probably not quite as intuitive as the word "connected." (What else could connected mean but "all in one piece"?) The word "compact" ought to mean something like "not too big and not too spread out," and, in topological language, should also mean not homeomorphic to a set that is "too big and too spread out." In the real line with its usual topology, let us agree that a set consisting of a single point should be compact (it is certainly "not too big"), and a closed and bounded interval should also be compact (being bounded, such an interval is "not too spread out"). But an open interval, even if it is bounded, should not be compact because such an interval is homeomorphic to **R** itself, which is as big and spread out as it is possible to be in **R**.

We want a definition of compactness, then, such that points and closed and bounded intervals will be compact subsets of **R** with its usual topology, while open intervals will not be compact. To get such a definition, let us investigate some of the properties of the closed unit interval [0, 1], a set that we certainly want to be compact.

The closed unit interval is a closed and bounded subset of **R** with its usual topology, which, in addition, satisfies the following.

1.1. Theorem. Let $\mathcal{U}$ be a collection of open subsets of **R** such that $[0, 1] \subset \bigcup \mathcal{U}$. Then there are finitely many sets in $\mathcal{U}$, $U_1, U_2, \cdots, U_n$, such that $[0, 1] \subset \bigcup_{i=1}^{n} U_i$.

[*Hint for proof*: Consider the set

$$\{x \in [0, 1]: [0, x] \text{ can be covered by finitely many of the sets in } \mathcal{U}\}.$$

Show that the supremum (least upper bound) of this set exists and is equal to 1.]

Depending on the source that you are reading, Theorem 1.1 is either called the Heine-Borel theorem[*] or is a corollary to a more general theorem which is called the Heine-Borel theorem. We will call Theorem 1.1 the Heine-Borel theorem and will prove the more general theorem later.

The Heine-Borel theorem says, then, that whenever [0, 1] is contained in the union of a collection of sets, then if this collection is infinite, almost all of the sets in the collection are not needed, because [0 1], is already completely contained in the union of only finitely many of the sets in the collection.

For a subset S of a topological space, whenever a collection of open sets $\mathcal{U}$ of the space is such that $S \subset \bigcup \mathcal{U}$, we call $\mathcal{U}$ an **open cover** of S, and any subcollection of $\mathcal{U}$ whose union also contains S is called an **open subcover** of S contained in $\mathcal{U}$. Thus the Heine-Borel theorem can be stated in words as follows:

Any open cover of [0, 1] contains a finite subcover.

Note that given an open cover of [0, 1], the finite subcover guaranteed by the Heine-Borel theorem consists of sets that belong to the original open cover; the subcover is a subcollection of the cover, and its members are

---

[*] After Eduard Heine (1821–1881) and Félix Edouard Emile Borel (1871–1956).

themselves *members* of the cover, rather than only subsets of members of the cover.

To reiterate, according to the Heine-Borel theorem, whenever [0, 1] is contained in the union of a collection of open sets, then actually only finitely many of these open sets are necessary to completely contain [0, 1] in their union. A theorem like the Heine-Borel theorem is *false* for the *open* unit interval, as you can show in Exercise 1.2 (1) below.

1.2. Exercises.

1) Give an example of an open cover of (0, 1) which does not contain a finite subcover. In other words, give an example of an infinite collection $\mathfrak{u}$ of open sets such that (0, 1) $\subset$ $\cup\mathfrak{u}$ and if $\mathfrak{u}'$ is any finite subcollection of $\mathfrak{u}$, (0, 1) $\not\subset \cup\mathfrak{u}'$.

2) Give an example of a finite subcover of [0, 1] from the open cover $\{[0, 1/(n+1)), (1/(n+1), 1], ((n-1)/2(n+1), (n+3)/2(n+1))\}$.

The closed unit interval is a closed and bounded subset of the real line and is such that every open cover of it contains a finite subcover of it. Since "bounded" is a term that is not always meaningful in a general topological space ("bounded" requires some kind of order on the space or some kind of distance function), we will base our definition of compactness in general on the open cover property of the closed unit interval, as stated in the Heine-Borel theorem.

1.3. Definition. A subset A of a topological space X is **compact** if every open cover of A contains a finite subcover.

Thus $A \subset X$ is compact if whenever $\mathfrak{u}$ is a collection of open subsets of X such that $A \subset \cup\mathfrak{u}$, then there is a *finite* subcollection $\mathfrak{u}'$ of $\mathfrak{u}$ such that A is also contained in $\cup\mathfrak{u}'$. In particular, a topological space itself is compact if every open cover of it contains a finite subcover.

1.4. Exercises.

1) State precisely what it means when a subset of a topological space is not compact.

2) What subsets of an indiscrete space are compact? In particular, is an indiscrete space itself ever compact?

3) What subsets of a discrete space are compact? In particular, is a discrete space itself ever compact?

4) What subsets of **R** with the finite complement topology are compact? In particular, is the space itself compact?

5) Is $\{1/n : n \in \mathbf{Z}^+\}$ a compact subset of **R** with its usual topology? How about $\{1/n : n \in \mathbf{Z}^+\} \cup \{0\}$?

6) Give an example of a closed subset of **R** with its usual topology (besides **R** itself) that is not compact.

7) Show that the open unit disk is not a compact subset of the plane with its usual topology. (The open unit disk is the region inside the unit circle centered at the origin but not containing any points of the circle.)

8) (This exercise requires a knowledge of ordinal numbers.) Prove that the space of countable ordinals, $[0, \Omega)$, is not compact, but $[0, \Omega]$ is compact. [*Hint:* To show that $[0, \Omega)$ is not compact is easy; to show that $[0, \Omega]$ is compact, use transfinite induction (Section 6 of Chapter 2).]

The following theorem is an easy consequence of the definition of compactness (if you remember that whenever A is a closed subset of a topological space X, then $X - A$ is open).

1.5. Theorem. A closed subset of a compact space is compact.

1.6. Exercise.

Show that a compact subset of a topological space need not be closed, even if the space itself is compact.

There is an important equivalent formulation of the definition of compactness in terms of closed sets and intersections rather than open sets and unions, which is an immediate consequence of the De Morgan laws and the definition of compactness.

1.7. Theorem. A topological space X is compact if and only if whenever $\mathfrak{F}$ is a collection of closed subsets of X such that the intersection of any finite subcollection of $\mathfrak{F}$ is not empty, then $\cap \mathfrak{F} \neq \emptyset$.

A collection of sets (like $\mathfrak{F}$ in Theorem 1.7) such that the intersection of any finite subcollection of the collection is not empty is said to have the **finite intersection property** (FIP). Thus Theorem 1.7 says that a topological space is compact if and only if every collection of *closed* sets with FIP has a non-empty intersection. A theorem like 1.7 can be given for a subset of a topological space by using the relative topology, as follows.

1.8. Theorem. A subset A of a topological space X is compact if and only if whenever $\mathfrak{F}$ is a collection of closed subsets of A (when A has the relative topology inherited from the topology on X) such that the intersection of any finite subcollection of $\mathfrak{F}$ is not empty, then $\cap \mathfrak{F} \neq \emptyset$.

1.9. Exercises.

1) Which of the following collections of subsets of **R** has FIP?

　a) $\{[-n, n] : n \in \mathbf{Z}^+\}$.

　b) $\{(-1/n, 1/n) : n \in \mathbf{Z}^+\}$.

　c) $\{[n, \infty) : n \in \mathbf{Z}^+\}$.

    d) $\{(n, \infty):n \in \mathbf{Z}^+\}$.
2) a) Prove, using the definition of compactness, that **R** with its usual topology is not compact.
    b) Prove, using Theorem 1.7, that **R** with its usual topology is not compact.
3) Prove that the Euclidean plane with its usual topology is not compact.
4) State precisely what it means when a set is not compact in terms of the finite intersection property.
5) The fact that the sets in Theorems 1.7 and 1.8 must be closed sets is essential. For example, we know that [0, 1] is compact. Give an example of a collection of subsets of [0, 1] with FIP such that the intersection of this collection is empty.
6) Prove that the set [0, 1] ∩ **Q** is not a compact subset of the space of rational numbers with the relative topology inherited from the usual topology on **R**.

Exercise 1.9 (6) above shows that compactness is a more "absolute" concept than closed is. The interval [0, 1] ∩ **Q** is the intersection of the closed and bounded (and therefore compact) subset [0, 1] of **R** with **Q**, and as the intersection of **Q** with a closed subset of **R**, is closed in **Q** with the relative topology inherited from the usual topology on **R**. But even though [0, 1] ∩ **Q** is the intersection of **Q** with the compact subset [0, 1] of **R**, [0, 1] ∩ **Q** is *not* a compact subset of **Q**. Thus "compact" does not behave as well with respect to the relative topology as "closed" does.

The reason for the "misbehavior" of compactness with respect to the relative topology is that a subset of a subspace of a space is closed if and only if it contains all of the points *of the subspace* that are close to it (the missing irrational numbers do not stop [0, 1] ∩ **Q** from being closed *in* **Q**, for example). On the other hand, a subset of a subspace of a space is compact if and only if it contains all points that are close to it even if these points are not in the space itself, let alone in the subspace. This is why a compact subset of **R** with its usual topology must be bounded: an unbounded subset of **R** is "close" to ∞ or − ∞ (or both), so cannot be compact because, as a subset of **R**, it cannot contain either of these points since they do not belong to **R**. An unbounded subset of **R** can be closed, however (indeed, **R** itself is closed in **R**), because closed is a relative term: to be closed in **R**, a set need only contain all points *of* **R** that are close to it. The points ∞ and − ∞, not being members of **R**, do not even enter into a discussion of whether or not a subset of **R** is closed, while, as mentioned above, they do enter into such a discussion concerning whether or not an unbounded subset of **R** is compact.

The discussion above makes it clear that to do much with compactness, we are going to have to have a precise definition of the idea of a point being close to a set. The idea of closure does this, but it is convenient to have the following definition as well. (This definition was given in Exercises 5.11 of Chapter 4; review it here if you did these exercises.)

1.10. Definition. Let X be a topological space and let $A \subset X$. A point $p \in X$ is a **cluster point** of A if every neighborhood of p meets A in at least one point other than p. The set of all cluster points of A is called the **derived set** of A and is denoted by A'.

Thus $p \in A'$ if and only if given any neighborhood N of p,

$$A \cap N - \{p\} \neq \emptyset.$$

1.11. Exercises.
   1) Let X be a topological space, $A \subset X$.
      a) Show that $\bar{A} = A \cup A'$.
      b) Is A necessarily equal to $\bar{A} - A'$?
   2) Let $X = \{a, b, c, d\}$ with the topology

$$\mathfrak{I} = \{\emptyset, X, \{a\}, \{a, b\}, \{c, d\}, \{a, c, d\}\}.$$

   What are $\{p\}'$ and $\overline{\{p\}}$ when $p = a, b, c,$ or d? What is $\{a, d\}'$? $\overline{\{a, d\}}$? What is $\{b, d\}'$? $\overline{\{b, d\}}$?
   3) Let X be a topological space, $A \subset X$. State precisely what it means when $p \in X$ is not a cluster point of A.

2. PRESERVATION OF COMPACTNESS BY CERTAIN OPERATIONS AND FUNCTIONS AND SOME CONSEQUENCES

To give us more examples to work with, we first establish the fact that the continuous image of a compact space is also compact.

2.1. Theorem. Let X and Y be topological spaces and let $f: X \to Y$ be continuous. Then if X is compact, $f(X)$ is also compact.

As a special case of Theorem 2.1, we have the fact that homeomorphic spaces are either both compact or both not compact, so a compact space cannot be homeomorphic to a space which is not compact. Also, a noncompact space cannot be a retract of a compact space. Thus, for example, $(0, 1)$ is not a retract of $[0, 1]$.

For a positive example of the importance of Theorem 2.1, we can prove the generalized form of the Heine-Borel theorem mentioned earlier. First, though, we observe that since any two closed and bounded intervals with more than one point are homeomorphic and since the closed unit interval is compact, we can prove the following.

2.2. Theorem. For $a, b \in \mathbf{R}$ with $a < b$, the interval $[a, b]$ is compact.

It is important to be aware of the fact that when **R** has its usual topology, even though all intervals of the form $[a, b]$ for $a < b$, a, $b \in \mathbf{R}$, are compact, there are compact subsets of **R** that are *not* intervals. Indeed, the generalized Heine-Borel theorem below (Theorem 2.3) completely characterizes compactness for subsets of the real line with its usual topology by saying that such a subset is compact if and only if it is closed and bounded, and certainly not every closed and bounded subset of **R** is a closed interval.

2.3 Theorem. A subset of the real line with its usual topology is compact if and only if it is closed and bounded.

[*Hint for proof:* For the "if" part, use Theorems 2.2 and 1.5 and recall the formal definition of a bounded subset of **R**: $A \subset \mathbf{R}$ is bounded if there exists a positive integer N such that $A \subset [-N, N]$. For the "only if" part, observe that if a subset A of **R** is not closed, then there exists a point in $\bar{A} - A$; use this point to violate the conclusion of Theorem 1.7. On the other hand, if $A \subset \mathbf{R}$ is not bounded, then either $A \cap [n, \infty) \neq \emptyset$ for all $n \in \mathbf{Z}^+$ or $A \cap (-\infty, -n] \neq \emptyset$ for all $n \in \mathbf{Z}^+$. In either case, Theorem 1.7 will show that A is not compact.]

There are two important corollaries to the generalized Heine-Borel theorem (Theorem 2.3). The first is the theorem mentioned in the introduction to this chapter, and the second is known as the **nested set property,** or, in a special case, the **nested interval property.**

2.4. Corollary. Let $f: C \subset \mathbf{R} \to \mathbf{R}$ be continuous, where C is a compact subset of **R** and **R** has its usual topology. Then there exist numbers x_M and x_m in C such that $f(x_M) = \max \{f(x): x \in C\}$ and $f(x_m) = \min \{f(x): x \in C\}$.

Thus a continuous real-valued function on a compact set in **R** attains both its maximum and minimum values. Notice how easily this follows from our previous work; you might recall how difficult it is to prove without general topological concepts. (The same might be said for the intermediate value theorem in Chapter 6.)

2.5. Corollary. (The nested set property.) If $S_1 \supset S_2 \supset S_3 \supset \cdots$ are non-empty closed and bounded subsets of **R** and **R** has its usual topology, then $\bigcap_{i=1}^{\infty} S_i \neq \emptyset$.

A sequence of sets $S_1, S_2, S_3, \cdots$ where each is contained in the preceding one is said to be **nested.** Thus a nested sequence of closed and bounded subsets of **R** has a non-empty intersection when **R** has its usual topology. In particular, if the nested sequence consists of closed and bounded intervals, we have the following.

2.6. Corollary. (The nested interval property.) If $I_1 \supset I_2 \supset I_3 \supset \cdots$ are non-empty closed and bounded intervals in **R** and **R** has its usual topology, then $\bigcap_{i=1}^{\infty} I_i \neq \emptyset$.

We investigate the preservation of compactness by certain set operations in the following exercises, in addition to observing that the hypotheses in some of the theorems in this section are necessary. We also look at some consequences of these theorems.

2.7. Exercises.

1) Must the union of compact subsets of a topological space be compact? State and prove a theorem concerning compactness and unions.

2) Surprisingly, the intersection of compact subsets of a space need not be compact. For example, consider the space X which is the set $[0, 1]$ with topology defined as follows: the only neighborhood of 0 and 1 is X itself, and for $x \in X$ with $x \neq 0$, $x \neq 1$, $\{x\}$ is open.
   a) Prove that the space described above is a topological space.
   b) Give an example of two compact subsets of X whose intersection is not compact.
   c) Give an example of a compact subset of X that is not closed.

3) It is the fact that the space in Problem 2 above contains compact subsets that are not closed that makes it such that the intersection of compact subsets need not be compact. Prove that if all compact subsets of a topological space are closed, then the intersection of compact subsets must be compact.

4) Must the complement of a compact subset of a topological space be compact? How about if the space itself is compact?

5) Show using the definition of compactness that any singleton subset of any topological space is compact. In particular, for $p \in \mathbf{R}$, $\{p\}$ is a compact subset of $\mathbf{R}$ with its usual topology. (This is also an immediate result of Theorem 2.3.)

6) Show using the definition of compactness that any finite subset of any topological space is compact. (For a finite subset of the real line with its usual topology, this is also an immediate result of theorem 2.3. For finite subsets of general topological spaces, it also follows from Problem 5 and your theorem in Problem 1.)

7) We know that $(0, 1) \cong \mathbf{R}$, both with their usual topologies. Prove that $[0, 1] \ncong \mathbf{R}$ when both have their usual topologies.

8) Prove that if the product of non-empty spaces is compact, then each of the factors must also be compact. (The converse is also true, but is much more difficult to show. We postpone it until we have more information to work with.)

9) By Corollary 2.4, a continuous function on a compact subset of $\mathbf{R}$ attains both its maximum and minimum values. In other words, if $f: C \subset \mathbf{R} \to \mathbf{R}$ where C is compact and $\mathbf{R}$ has its usual topology, then there are numbers x_M and x_m in C such that

$$f(x_M) = \sup \{f(x): x \in C\}$$

and

$$f(x_m) = \inf \{f(x): x \in C\}.$$

a) Give an example of a bounded continuous real-valued function on (0, 1) which does not attain either its maximum or its minimum. (A function $f: (0, 1) \to \mathbf{R}$ is *bounded* if $f(0, 1)$ is a bounded subset of **R**.)

b) Give an example of a real-valued function on [0, 1] that does not attain its maximum or minimum. Can this function be continuous?

10) By the nested set property, a nested sequence of closed and bounded subsets of **R** has a non-empty intersection. Give an example of a sequence of subsets of **R** whose intersection is empty. Can your sets be closed? Can they be bounded?

3. SOME EQUIVALENCES OF COMPACTNESS IN R

Throughout this section, let **R** have its usual topology and let all subsets of **R** have the relative topology inherited from the usual topology on **R**.

To summarize what we have so far, a subset of a topological space is compact if every open cover of it has a finite subcover, and in **R**, compactness of a set is equivalent to its being closed and bounded. There are several other ways to characterize compactness for subsets of **R**, which we examine now.

Recall the definition of cluster point given in Section 1. The following theorem, know as the Bolzano-Weierstrass theorem,* is a classical theorem of analysis.

3.1. Theorem. A subset A of **R** is compact if and only if every infinite subset of A has a cluster point in A.

[*Hint for proof:* Use the "closed and bounded" definition of compactness. For the "if" part, suppose that A is not compact, so that it is not closed or not bounded. In either case, you can construct an infinite subset of A with no cluster point in A. For the "only if" part, let $S \cap A$ be infinite, where A is compact, so is contained in an interval of the form $[-n, n]$ for some $n \in \mathbf{Z}^+$. Cut this interval in half and observe that the part of S in one half or the other (or both) must be infinite. Choose a half in which S is infinite and cut it in half. Continue this process and use the nested interval property to find a cluster point of S.]

* After Bernhard Bolzano (1781–1848) and Karl Theodor Wilhelm Weierstrass (1815–1897).

It is important to notice that the infinite set in the Bolzano-Weierstrass theorem need only be *countably* infinite, as you can show in Exercise 3.2 (1) below.

3.2. Exercises.

1) Show that a subset A of **R** is compact if and only if every countably infinite subset of A has a cluster point in A. [*Hint:* Read the definition of cluster point very carefully. Is a cluster point of a countably infinite subset of a set also a cluster point of the set itself? Of course you can use the Bolzano-Weierstrass theorem if you need it.]

2) State precisely what it means when a subset A of **R** is not compact, according to the Bolzano-Weierstrass theorem.

3) Give an example of an infinite subset of **R** with no cluster point in **R**, thus showing that **R** is not compact by the Bolzano-Weierstrass theorem.

4) Give an example of an infinite subset of $(0, 1)$ with no cluster point in $(0, 1)$, thus showing that $(0, 1)$ is not compact by the Bolzano-Weierstrass theorem.

5) Prove using the Bolzano-Weierstrass theorem that $[0, 1] \cap \mathbf{Q}$ is not a compact subset of **Q**. Is $[0, 1] \cap \mathbf{Q}$ closed in **Q**?

6) The Bolzano-Weierstrass theorem says that a subset A of **R** is compact if and only if every infinite subset of A has a cluster point *in A*. The "in A" is important. Show that a subset A of **R** is bounded if and only if every infinite subset of A has a cluster point in $\bar{A} \subset \mathbf{R}$.

    Thus when A is compact, it is closed as well as bounded, so the cluster point guaranteed by the fact that A is bounded must be in A because A is closed. On the other hand, if every infinite subset of A has a cluster point in A, then A must be bounded simply because such cluster points exist *in* **R**, and must be closed because they are all *in* A; in other words, A is compact.

Another way to look at compactness in **R**, which is very much like the Bolzano-Weierstrass theorem, involves using sequences. Recall (Theorem 5.5 in Chapter 3) that a subset A of **R** is closed if and only if whenever a sequence contained in A converges to a point $x \in \mathbf{R}$, then $x \in A$. Notice that this requires that the sequence in question already converges to a point, and says nothing about the behavior of sequences that do not converge. Even compactness cannot force a non-convergent sequence to converge, but it does require that *every* sequence in a compact subset of **R**, whether it converges or not, contain a sequence that does converge. To discuss this further, we need to make precise the idea of a sequence contained in a sequence. Basically this means just what it says, but some care must be taken so that a *subsequence* of a sequence has the desired properties.

3.3. Definition. Let $\{x_n : n \in \mathbf{Z}^+\}$ be a sequence. A sequence $\{x_{n_k} : k \in \mathbf{Z}^+\}$ is a **subsequence** of $\{x_n : n \in \mathbf{Z}^+\}$ if

1) $n_{k+1} > n_k$ for each $k \in \mathbf{Z}^+$.
2) for each $k \in \mathbf{Z}^+$, there exists an $n \in \mathbf{Z}^+$ such that $n_k = n$ (so $x_{n_k} = x_n$).
3) $n_k \geq k$ for each $k \in \mathbf{Z}^+$.
4) for each $n \in \mathbf{Z}^+$, there exists a $k \in \mathbf{Z}^+$ such that $n_k \geq n$.

Thus a sequence $\{y_m : m \in \mathbf{Z}^+\}$ is a subsequence of $\{x_n : n \in \mathbf{Z}^+\}$ if the terms of $\{y_m : m \in \mathbf{Z}^+\}$ appear frequently in $\{x_n : n \in \mathbf{Z}^+\}$, in order. This requires more than just that the point-set determined by a subsequence be a subset of the point set determined by the sequence. A subsequence of a sequence is really a subset *as a sequence*, so the ordering of the terms induced by the positive integers must be taken into account. A subsequence of a sequence is a subset, but must be a subset infinitely often as n goes to ∞, and, furthermore, cannot double back on itself. Some examples will help to make this clear.

3.4. Exercises.

1) Which of the following sequences is a subsequence of the sequence $\{1, 2, 3, \cdots\} = \{x_n : n \in \mathbf{Z}^+\}$ where $x_n = n$. Which are merely subsets of the point-set determined by this sequence?

   a) $\{1, 4, 9, \cdots\} = \{y_m : m \in \mathbf{Z}^+\}$ where $y_m = m^2$.
   b) $\{1, 2, 1, 2, \cdots\} = \{y_m : m \in \mathbf{Z}^+\}$ where

   $$y_m = \begin{cases} 1, & \text{if } m \text{ is odd} \\ 2, & \text{if } m \text{ is even.} \end{cases}$$

   c) $\{2, 3, 5, 7, 11, 13, \cdots\} = \{y_m : m \in \mathbf{Z}^+\}$ where y_m is the m-th prime.
   d) $\{1, 2, 1, 4, 5, 6, 7, 8, 1, 10, \cdots\} = \{y_m : m \in \mathbf{Z}^+\}$ where

   $$y_m = \begin{cases} 1, & \text{if } m = 3^n \text{ for some } n \in \mathbf{Z}^+. \\ m, & \text{otherwise} \end{cases}$$

2) a) Give an example of a convergent subsequence of a non-convergent sequence in $\mathbf{R}$.
   b) Can a convergent sequence in $\mathbf{R}$ have a non-convergent subsequence?
   c) Can a sequence in $\mathbf{R}$ have convergent subsequences that converge to different points?
   d) Can a convergent sequence in $\mathbf{R}$ have a subsequence that converges to a different point than the sequence does?

3) Recall that a sequence in X is a function from $\mathbf{Z}^+$ into X. We can also define a subsequence as a function. To do it, we need a defini-

tion: if $A, B \subset \mathbf{R}$, a function $f: A \rightarrow B$ is *monotone increasing* if whenever $a_1 < a_2$ in A, then $f(a_1) < f(a_2)$ in B.

Show that $T: \mathbf{Z}^+ \rightarrow X$ is a subsequence of $S: \mathbf{Z}^+ \rightarrow X$ if and only if there exists a monotone increasing function $F: \mathbf{Z}^+ \rightarrow \mathbf{Z}^+$ such that $T = S \circ F$. [*Hint:* For $k \in \mathbf{Z}^+$, write $F(k) = n_k$.]

Before getting a characterization of compactness for subsets of $\mathbf{R}$ in terms of sequences and subsequences, we establish the following useful lemma which gives a way to get convergent subsequences out of certain sequences. Recall that a point is an accumulation point of a sequence if the sequence is frequently in every neighborhood of the point, and a sequence converges to a point if the sequence is ultimately in every neighborhood of the point.

3.5. Lemma. A point x_0 in $\mathbf{R}$ with its usual topology is an accumulation point of a sequence of points $\{x_n : n \in \mathbf{Z}^+\} \subset \mathbf{R}$ if and only if there is a subsequence of $\{x_n : n \in \mathbf{Z}^+\}$ that converges to x_0.

Some confusion may be caused by the similarities and differences between a *cluster point* of a *set* of points and an *accumulation point* of a *sequence* of points. The following exercises may help to clarify things.

3.6. Exercises.
1) Give an example to show that an accumulation point of a sequence $\{x_n : n \in \mathbf{Z}^+\}$ in $\mathbf{R}$ need *not* be a cluster point of the set of points $\{x_n : n \in \mathbf{Z}^+\} \subset \mathbf{R}$. [*Hint:* There are infinitely many $n \in \mathbf{Z}^+$ but there may be only finitely many distinct points x_n.]
2) Show that if $\{x_n : n \in \mathbf{Z}^+\}$ is a sequence in $\mathbf{R}$ with all x_n distinct, then an accumulation point of the sequence $\{x_n : n \in \mathbf{Z}^+\}$ is a cluster point of the set of points $\{x_n : n \in \mathbf{Z}^+\} \subset \mathbf{R}$.
3) Show that a cluster point of the set of points $\{x_n : n \in \mathbf{Z}^+\} \subset \mathbf{R}$ is an accumulation point of the sequence $\{x_n : n \in \mathbf{Z}^+\}$ of points of $\mathbf{R}$. [*Hint:* That a cluster point of $\{x_n : n \in \mathbf{Z}^+\}$ exists implies that infinitely many of the points are different.]
4) Are the following statements true or false:
   a) A cluster point of the set of points $\{x_n : n \in \mathbf{Z}^+\} \subset \mathbf{R}$ is an accumulation point of the sequence $\{x_n : n \in \mathbf{Z}^+\}$ in $\mathbf{R}$ if and only if there are infinitely many distinct points x_n.
   b) An accumulation point of the sequence $\{x_n : n \in \mathbf{Z}^+\}$ in $\mathbf{R}$ is a cluster point of the set of points $\{x_n : n \in \mathbf{Z}^+\} \subset \mathbf{R}$ if and only if there are infinitely many distinct points x_n.

We can now get a characterization of compactness in $\mathbf{R}$ in terms of sequences and subsequences.

3.7. Theorem. A subset A of **R** is compact if and only if every sequence in A has a convergent subsequence that converges to a point of A.
[*Hint for proof:* Use the Bolzano-Weierstrass theorem.]

3.8. Exercises.

1) State precisely what it means according to Theorem 3.7 when a subset of **R** is not compact.

2) Give an example of a sequence in **R** with no convergent subsequence, thus showing that **R** is not compact according to Theorem 3.7.

3) Give an example of a sequence in $(0, 1)$ with no convergent subsequence, thus showing that $(0, 1)$ is not compact according to Theorem 3.7.

4) Theorem 3.7 says that a subset A of **R** is compact if and only if every sequence in A has a convergent subsequence that converges to a point *in A*. The "in A" is important.

   Show that a subset A of **R** is bounded if and only if every sequence in A has a convergent subsequence that converges to a point in **R**.

   Thus, when A is compact, it is closed as well as bounded, so every sequence in A has a convergent subsequence because A is bounded, and, since A is closed, this subsequence must converge to a point in A. On the other hand, if every sequence in A has a convergent subsequence that converges to a point in A, then A is bounded simply because such subsequences exist, and is closed because they must converge to points in A; in other words, A is compact.

We can get another characterization of compactness for a subset A of **R** in terms of the continuous real-valued functions defined on A. First, we repeat the following definition.

3.9. Definition. Let X be a set. A function $f: X \to$ **R** is said to be **bounded** if $f(X)$ is a bounded subset of **R**, i.e., if there exists a positive integer N such that $f(X) \subset [-N, N]$.

Notice that in the definition of a bounded function, nothing is said in any way about $f(X)$ being closed, so a bounded real-valued function must have a *bounded* image in **R**, but not necessarily a *compact* image. But the continuous image of a compact set is compact, so, in particular, a continuous real-valued function on a compact subset of **R** is bounded (i.e., is a bounded function). We will use this property to get another characterization of compactness for subsets of **R**. To do it, we first establish some more machinery to work with.

Recall that if we have a continuous function defined on a topological space X with range in a topological space Y, then the restriction of f to A, $f \mid A : A \to Y$, is also continuous when A has the relative topology inherited from the topology on X, where $(f \mid A)(a) = f(a)$ for all $a \in A$. It is often (but not always!) possible to turn this situation around and to *extend* a continuous function from $A \subset X$ into Y to a continuous function from all of X into Y, such that the new function gives the same values as the old one to points in A. The question of when such continuous extensions exist is a very important problem in topology. To see that it *is* a problem, consider Exercise 3.10 (1) below.

3.10. Exercises.

1) Consider the continuous function $f : (0, 1] \to \mathbf{R}$ defined by $f(x) = \sin(1/x)$. Show that it is *impossible* to extend this function to a continuous function from $[0, 1]$ into $\mathbf{R}$. In other words, show that there does not exist a continuous function $F : [0, 1] \to \mathbf{R}$ such that $f \mid (0, 1] = f$ (i.e., such that $F(x) = \sin(1/x)$ for $0 < x \leq 1$).

2) Let A be a retract of a topological space X. Show that if $f : A \to Y$ is continuous, where Y is a topological space, then there is a continuous extension of f to all of X. (In other words, there is a continuous function $F : X \to Y$ such that $F \mid A = f$.)

3) Is the converse to Problem 2 true?

Later on we will consider the problem of extending functions in some generality, but for now let us establish the following (very) special case. Recall the definition of a monotone increasing function. Since a sequence is a function (whose domain is the set of positive integers), this definition applies to sequences as well, and we can say that a sequence is **monotone increasing** if $x_n < x_{n+1}$ for all $n \in \mathbf{Z}^+$, and is **monotone decreasing** if $x_n > x_{n+1}$ for all n. Note that $\{x_n : n \in \mathbf{Z}^+\}$ is a discrete subspace of $\mathbf{R}$, in either case.

3.11. Lemma. Let $\{x_n : n \in \mathbf{Z}^+\}$ be a sequence in $A \subset \mathbf{R}$ which is monotone increasing or monotone decreasing, and let $f : \{x_n : n \in \mathbf{Z}^+\} \to \mathbf{R}$ by $f(x_n) = n$ for each n. Then f is continuous on $\{x_n : n \in \mathbf{Z}^+\}$, and if this set is a closed subset of A, f can be extended to a continuous function $F : A \to \mathbf{R}$. [*Hint for proof:* Suppose for the moment that $\{x_n : n \in \mathbf{Z}^+\}$ is monotone increasing. The intervals (x_n, x_{n+1}) and $(n, n + 1)$ are very much alike. Try patching together some restrictions of homeomorphisms to get an extension of f. What are you going to do with the part of A below x_1, $\{x \in A : x < x_1\}$?]

When $\{x_n : n \in \mathbf{Z}^+\} \subset A \subset \mathbf{R}$ is monotone increasing and, in addition, converges to a point $x_0 \in \mathbf{R}$, we write $x_n \uparrow x_0$; the symbol $x_n \downarrow x_0$ is defined analogously.

3.12. Lemma. If x_0 is a cluster point of $A \subset \mathbf{R}$, then there exists a sequence $\{x_n : n \in \mathbf{Z}^+\} \subset A$ such that either $x_n \uparrow x_0$ or $x_n \downarrow x_0$.

Putting Lemmas 3.11 and 3.12 together with the Bolzano-Weierstrass theorem, you can prove the following characterization of compactness of subsets of **R**.

3.13. Theorem. A subset A of **R** is compact if and only if every real-valued continuous function defined on A is bounded.

3.14. Exercises.
1) Give an example of a non-constant bounded function on $(0, 1)$.
2) Give an example of a non-constant bounded function on **R**.
3) Is every continuous real-valued function on $[0, 1] \cap \mathbf{Q}$ bounded?
4) State precisely what it means when $A \subset \mathbf{R}$ is not compact according to Theorem 3.13.
5) The distance between two subsets A and B of a metric space (X, d) is defined to be inf $\{d(a, b) : a \in A, b \in B\}$.
   a) Prove that if A and B are compact subsets of **R** with its usual metric, then if A and B are disjoint, $d(A, B) > 0$.
   b) Give an example of two disjoint closed subsets of **R** with its usual metric such that the distance between these two sets is 0. [*Hint:* By (a), the sets cannot both be compact. Can one of them be compact?]
7) The fact that the sequence in Lemma 3.11 must be a closed subset of A is essential. For example, let $A = [-1, 0]$ and define

$$f : \{-1/n : n \in \mathbf{Z}^+\} \to \mathbf{R} \text{ by } f(-1/n) = n.$$

Show that f cannot be extended to a continuous function defined on all of A.
8) The function in Problem 7 that cannot be extended from

$$\{-1/n : n \in \mathbf{Z}^+\} \text{ to } [-1, 0]$$

is not bounded. Can you give an example of such a function that is bounded?

VARIATIONS OF COMPACTNESS

We collect the results that we have so far in the following theorem.

4.1. Theorem. For $A \subset \mathbf{R}$ with its usual topology, the following are equivalent:
1) A is compact; i.e., every open cover of A has a finite subcover.
2) A is closed and bounded.

3) Every infinite subset of A has a cluster point in A.
4) Every sequence in A has a convergent subsequence (that converges to a point in A).
5) Every real-valued continuous function on A is bounded.

In a general topological space, none of these concepts is equivalent to any other. In this section, we will see examples of this, and will also see that the definition of compactness (namely that every open cover has a finite subcover) is a good one, because, even in a general topological space, when a set is compact, it also satisfies (3) and (5) of Theorem 4.1, and in metric spaces, a compact set satisfies all five statements in the theorem.

Conditions (3), (4), and (5) in Theorem 4.1 are given names as follows.

4.2. Definition. Let X be a topological space and let A be a subspace of X (i.e., A has the relative topology inherited from the topology on X).
1) A is **countably compact** if every infinite subset of A has a cluster point in A.
2) A is **sequentially compact** if every sequence in A has a convergent subsequence (that converges to a point in A).
3) A is **pseudocompact** if every real-valued continuous function defined on A is bounded.

As with the Bolzano-Weierstrass theorem in **R,** the following simple lemma often makes countable compactness easier to work with, and also justifies the name (although we will see a better justification of the name later).

4.3. Lemma. A subspace A of a topological space X is countably compact if and only if every countably infinite subset of A has a cluster point in A.

4.4. Theorem. Let X be a topological space and let A be a compact subset of X. Then A is also both countably compact and pseudocompact. [*Hint for proof:* To show that A is pseudocompact is easy with what we know now; to show that A is countably compact, show that if A contains an infinite set with no cluster point then there is an open cover of A with no finite subcover. To construct such an open cover, recall what it means for a point not to be a cluster point of a set.]

Theorem 4.4 cannot be strengthened. In other words, there are countably compact sets which are not compact, and there are pseudocompact sets which are not compact. Also, there are pseudocompact sets which are not countably compact, but every countably compact set must also be pseudocompact. You can show all of this in the following exercises.

4.5. Exercises.

1) A countably compact space need not be compact.
   a) (Using ordinal numbers.) The space of countable ordinal numbers $[0, \Omega)$ is countably compact but not compact. [*Hint:* See Section 5 in Chapter 2.]
   b) (Not using ordinal numbers.) Let $X = \bigcup_{n=1}^{\infty} (n, n + 1)$ with topology $\mathfrak{I}$ generated by the basis $\{(n, n + 1): n \in \mathbf{Z}^+\}$. Prove that $(X, \mathfrak{I})$ is a countably compact space that is not compact. [*Hint:* This topology is *nothing* like the usual topology on X as a subspace of **R**.]

2) Every countably compact space is also pseudocompact. [*Hint:* See the proof of Theorem 3.13.]

3) (This exercise requires a knowledge of ordinal numbers.) From Problems 1 and 2, it is immediate that the space of countable ordinals is pseudocompact. In fact, it is actually more than just pseudocompact because a continuous real-valued function on $[0, \Omega)$ is ultimately constant, so is more than just bounded. (A function f on $[0, \Omega)$ is *ultimately constant* if there is an ordinal number $a \in [0, \Omega)$ such that $f(x) = f(y)$ for all $x, y \in (a, \Omega)$.) Showing that every continuous real-valued function on $[0, \Omega)$ is ultimately constant is not easy. Use the following outline, which is adapted from Chapter 5 of Gillman and Jerison [6].
   a) If A and B are closed subsets of $[0, \Omega)$, then at least one of A or B is bounded (i.e., there is an ordinal number $a < \Omega$ such that $A \subset [0, a]$ or $B \subset [0, a]$.
   b) Show that for each ordinal number $a \in [0, \Omega)$, $[a, \Omega)$ is countably compact.
   c) Show that the continuous image of a countably compact space is countably compact.
   d) Let $f: [0, \Omega) \to \mathbf{R}$ be continuous and show that for each ordinal number $a \in [0, \Omega)$, $f(a, \Omega)$ is a *compact* subset of **R** (see Theorem 4.1). Deduce that $\bigcap\{f(a, \Omega): a \in [0, \Omega)\} \subset \mathbf{R}$ is not empty, and choose a real number r in this intersection. Show that for each $n \in \mathbf{Z}^+$, the set $\{x \in [0, \Omega): f(x) \geq 1/n\} = S_n$ is closed in $[0, \Omega)$ and is disjoint from $f^{-1}(\{r\})$, which is also closed in $[0, \Omega)$. Using (a) and the properties of r, show that S_n is bounded and let $a_n \in [0, \Omega)$ be such that $S_n \subset [0, a_n)$. Use Section 5 of Chapter 2 to get an ordinal number $a \in [0, \Omega)$ such that $a_n \leq a$ for all $n \in \mathbf{Z}^+$, and show that $f(x) = r$ for all $x \in (a, \Omega)$. Conclude that f is ultimately constant, so every real-valued continuous function on $[0, \Omega)$ is ultimately constant. In particular, $[0, \Omega)$ is pseudocompact.

4) A pseudocompact space need not be countably compact. Let $X = \mathbf{R}$

with the *countable complement* topology defined by $U \subset X$ is open if and only if $U = \emptyset$ or $X - U$ is countable. To see that X is not countably compact is easy. To show that it is pseudocompact, show that every real-valued continuous function defined on X is constant, and therefore is bounded.

To summarize what we have so far (using "$\Rightarrow$" to denote "implies"):

$$\text{compact} \Rightarrow \text{countably compact} \Rightarrow \text{pseudocompact},$$

and none of the implications is reversible in general (although in **R** with its usual topology, all three concepts are the same.)

Before putting the other two compactness ideas (closed and bounded, and sequentially compact) on the diagram, we establish the fact that both countable compactness and pseudocompactness are preserved by continuous functions.

4.6. Theorem. Let X and Y be topological spaces and let $f: X \to Y$ be continuous. Then if X is countably compact, $f(X)$ is also countably compact, and if X is pseudocompact, $f(X)$ is also pseudocompact.

4.7. Exercises.
1) In a metric space, a compact set must be closed and bounded. (In a general metric space (X, d), a set A is bounded if there is a real number r such that $d(x, y) < r$ for all $x, y \in A$. For a non-empty bounded set A in (X, d), the real number $\sup\{d(x, y) : x, y \in A\}$ is called the **diameter** of A, and is denoted by $\delta(A)$.) [*Hint for proof:* Show that a countably compact subset of a metric space must be closed and bounded.]
2) In a metric space, a closed and bounded set need not be compact.

Thus, according to the exercises above, a compact subset of a metric space must be closed and bounded, but a closed and bounded subset of a metric space need not be compact. In **R** with its usual topology, however, a set is compact if and only if it is closed and bounded, as we have seen.

That $[0, 1] \cap \mathbf{Q}$, for example, which is a closed and bounded subset of $\mathbf{Q}$ with its usual topology, is not compact is because of its many missing points (namely, the irrational numbers between 0 and 1). Intuition about compactness and missing points has to be used very carefully, though, as the following very important example shows.

4.8. Exercise. (The Cantor set*.)
The **Cantor set** C is defined to be the subset of $[0, 1]$ obtained by

---

* After Georg Cantor (1845–1918), who did much work on set theory and developed the theory of infinite sets.

"deleting middle thirds," as follows: from [0, 1], remove all points in ($\frac{1}{3}$, $\frac{2}{3}$), all points in ($\frac{1}{9}$, $\frac{2}{9}$) and in ($\frac{7}{9}$, $\frac{8}{9}$), all points in ($\frac{1}{27}$, $\frac{2}{27}$), ($\frac{7}{27}$, $\frac{8}{27}$), ($\frac{19}{27}$, $\frac{20}{27}$) and in ($\frac{25}{27}$, $\frac{26}{27}$), etc. The first few stages in this deletion process are illustrated below.

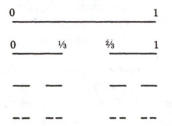

The **ternary decimal expansion** of a real number is a decimal representation of the number in base 3. (Our usual decimal representation of a number is in base 10.) Any real number r between 0 and 1 has a ternary decimal expansion of the form $r = 0.d_1d_2d_3 \cdots$, where each d_i is either 0, 1, or 2, and we can write this expansion as

$$r = d_1/3 + d_2/9 + d_3/27 + \cdots .$$

(Thinking of ordinary base 10 decimals should help to clarify this idea.)

From the summation expression of the ternary decimal expansion of a number between 0 and 1, it should be clear that a number $r = 0.d_1d_2d_3 \cdots$ having $d_1 = 1$ is in [$\frac{1}{3}$, $\frac{2}{3}$], a number having $d_1 \neq 1$ and $d_2 = 1$ is in either [$\frac{1}{9}$, $\frac{2}{9}$] or [$\frac{7}{9}$, $\frac{8}{9}$], etc. Noting that $\frac{1}{3}$, for example, can be written in a ternary expansion as either 0.1000 $\cdots$ or as 0.0222 $\cdots$ (just like 0.1000 $\cdots$ = 0.0999 $\cdots$ in base 10 decimals), it should not be too hard to convince yourself that the Cantor set consists precisely of those points that can be written in a ternary decimal expansion as $0.d_1d_2d_3 \cdots$ with no $d_i = 1$.

1) Prove the following, where **R** has its usual topology.
   a) The Cantor set is totally disconnected (i.e., its components are singleton sets).
   b) The Cantor set is a closed subset of **R**. [*Hint:* The intersection of closed sets is closed.]
   c) The Cantor set is a bounded subset of **R**.
   d) The Cantor set is a compact subset of **R**.

Thus [0, 1] ∩ **Q** is a totally disconnected bounded subset of **R** which is *not* compact, while the Cantor set is a totally disconnected bounded subset of **R** which *is* compact, so having infinitely many holes arbitrarily close together does not necessarily mean that a space is not compact.

2) (This exercise requires a knowledge of infinite products.) Prove that

the Cantor set C is homeomorphic to the product space $P = \prod_{i=1}^{\infty} X_i$ where each $X_i = \{0, 2\}$ with the discrete topology. [*Hint:* A point in the product is an ordered tuple $(d_1, d_2, d_3, \cdots)$ where each $d_i = 0$ or 2, and a point in the Cantor set has the form $0.d_1 d_2 d_3 \cdots$ where each $d_i = 0$ or 2; a 1-1 function f from C onto P should be clear. To show that this function is continuous in both directions (as it must be to be a homeomorphism), observe that C, being a subspace of $\mathbf{R}$, is a metric space so has a neighborhood basis consisting of r-balls. For any $r > 0$ (no matter how small), since $\sum_{i=1}^{\infty} d_i/3^i$, where $d_i = 0$ or 2, converges (indeed, this sum is a point in the Cantor set), then there is a positive integer N such that $\sum_{i=N}^{\infty} d_i/3^i < r$. Use a straightforward continuity argument.]

a) Prove that the Cantor set is not a discrete subspace of $\mathbf{R}$. [*Hint:* Any space homeomorphic to a discrete space must also be discrete.]

b) Prove that the product space $\prod_{i=1}^{\infty} X_i$, where each $X_i = \{0, 2\}$ with the discrete topology, is compact.

c) Prove that $\prod_{i=1}^{\infty} X_i$, where each $X_i = \{0, 2\}$ with the discrete topology, is *not* compact when it is given the box topology in which the product of open sets is open.

Exercises 4.8 (2) (b) and (c) are a good argument for defining the product topology as we did. With the product topology, the product of compact spaces is compact (Exercise 4.8 (2) (b) is an example of this; we will prove it in general later). But with the box topology, the product of compact spaces need not be compact, as Exercise 4.8 (2) (c) shows.

To return to our study of variations of compactness, we have so far that

$$\text{compact} \Rightarrow \text{countably compact} \Rightarrow \text{pseudocompact}$$
$$\searrow \text{(in metric spaces)}$$
$$\text{closed and bounded}$$

and none of these implications is reversible in general (although all of these concepts are equivalent in $\mathbf{R}$ with its usual topology).

To put sequential compactness on the diagram, prove the following simple theorem.

4.9. Theorem. Every sequentially compact space is countably compact.

The surprising thing about sequential compactness is that a compact space need *not* be sequentially compact, unlike the other variations of compactness that we have seen. Also, a sequentially compact space need not be compact, so there is absolutely no relation between compactness and sequential compactness in general.

4.10. Exercise.

(This exercise requires a knowledge of ordinal numbers.) Show that the space of countable ordinals is sequentially compact but not compact.

We will have to postpone showing that a compact space need not be sequentially compact until we know that the product of compact spaces is compact. We can show now that a theorem like Lemma 3.5, which says that a point in **R** is an accumulation point of a sequence of points of **R** if and only if there is a subsequence of the sequence that converges to the point, is false in general.

4.11. Exercise.*

Let X be the set of points in the plane with positive integer coordinates and define a topology on X as follows: each point in X except $(0, 0)$ is open, and open neighborhoods of $(0, 0)$ are such that an open set U containing $(0, 0)$ contains all but finitely many points of X on each of all but finitely many of the "lines" $\{n\} \times \mathbf{Z}^+ \subset X$. In symbols, $U \subset X$ containing $(0, 0)$ is open if and only if for all but at most finitely many $m \in \mathbf{Z}^+$, the set $\{n \in \mathbf{Z}^+ : (m, n) \notin U\}$ is finite. Let $\{x_n : n \in \mathbf{Z}^+\}$ be the sequence in X obtained by "counting the squares" as we did in Theorem 1.7 in Chapter 2. Thus $\{x_n : n \in \mathbf{Z}^+\}$ is the sequence $\{(1, 1), (1, 2), (2, 1), (3, 1), (2, 2), (1, 3), (1, 4), (2, 3), (3, 2), (4, 1), \cdots\}$.

1) Prove that $(0, 0)$ is an accumulation point of $\{x_n : n \in \mathbf{Z}^+\}$, but
2) Prove that no subsequence of $\{x_n : n \in \mathbf{Z}^+\}$ can converge to $(0, 0)$.

Thus even if this space were compact (which it is not), it could not be sequentially compact. When you think about it, the fact that a sequence can accumulate at a point while having no subsequence that converges to the point is *very* strange behavior indeed.

Although compactness and sequential compactness are not related in general, these two concepts are actually equivalent in a very important class of spaces, namely, metric spaces. Since we know that every compact space is countably compact, showing that compactness implies sequential compactness in metric spaces is easy because it is not hard to show that countable compactness implies sequential compactness in metric spaces. The key to the proof is the fact that every metric space (X, d) is 1st countable: for $x \in X$, the collection $\{S_{1/n}(x) : n \in \mathbf{Z}^+\}$ is a countable local basis at x.

* This example is due to R. Arens and M. K. Fort, Jr.

4.12. Theorem. A compact subset of a metric space is also sequentially compact.

To reverse the implication and show that sequential compactness implies compactness in metric spaces, we first introduce a new concept. Recall that the *diameter* of a non-empty subset A of a metric space (X, d) is defined to be

$$\delta(A) = \sup \{d(x, y) : x, y \in A\},$$

(which is a real number if A is bounded, and is $+\infty$ if A is not bounded).

4.13. Definition. Let $\mathfrak{U} = \{U_\lambda : \lambda \in \Lambda\}$ be an open cover of a metric space (X, d). A real number $r > 0$ is called a **Lebesgue number*** for $\mathfrak{U}$ if whenever $A \subset X$ with $\delta(A) < r$, then there is a $U_\lambda \in \mathfrak{U}$ such that $A \subset U_\lambda$.

Not every open cover of every metric space has a Lebesgue number, as you can show in the following exercise.

4.14. Exercise.

Let (X, d) be $(0, 1)$ with the usual metric and let $\mathfrak{U}$ be the open cover of X defined by

$$\mathfrak{U} = \{(\tfrac{1}{3}, 1), (\tfrac{1}{4}, \tfrac{1}{2}), (\tfrac{1}{5}, \tfrac{1}{3}), (\tfrac{1}{6}, \tfrac{1}{4}), \cdots \}$$
$$= \{(1/(n + 2), 1/n) : n \in \mathbf{Z}^+\}.$$

Show that there is no Lebesgue number for this open cover.

When (X, d) is a *compact* metric space, the behavior shown in Exercise 4.14 cannot happen, as you can show in the following theorem.

4.15. Theorem. Every open cover of a compact metric space has a Lebesgue number.

Outline of Proof: Let $\mathfrak{U} = \{U_\lambda : \lambda \in \Lambda\}$ be an open cover of X. Show that for each $x \in X$, there is a positive real number $r(x)$ such that $S_{r(x)}(x)$ is contained in some U_λ and that $\mathcal{S} = \{S_{r(x)/2}(x) : x \in X\}$ is an open cover of X, so has a finite subcover $\mathcal{S}'$. Let $r = \inf \{r(x)/2 : S_{r(x)/2}(x) \in \mathcal{S}'\}$ and show (using the triangle inequality for d) that r is a Lebesgue number for $\mathfrak{U}$.

It is also true that every open cover of a sequentially compact metric space has a Lebesgue number, and this is the property that we will use to show that sequential compactness implies compactness in metric spaces. This method is adapted from Simmons [14].

4.16. Theorem. Every open cover of a sequentially compact metric space has a Lebesgue number.

Outline of proof: Let $\mathfrak{U}$ be an open cover of the sequentially compact metric space (X, d) and let $\mathcal{S}$ be the collection of all subsets of X that are

---

* After Henri Lebesgue (1875–1941) who did much work in analysis.

not contained in any member of $\mathcal{U}$. Let $s = \inf\{\delta(S) : S \in \mathcal{S}\}$. If $s > 0$ or s "$= \infty$," the rest of the proof is easy (why?). Since s cannot be negative (why not?), the only other possibility is $s = 0$, a case that we can show is impossible. Indeed, if $s = 0$, then for each $n \in \mathbf{Z}^+$, there is an $S_n \in \mathcal{S}$ with $\delta(S_n) < 1/n$. Construct a sequence $\{x_n : n \in \mathbf{Z}^+\} \subset X$ by choosing $x_n \in S_n$, and finish the proof by using the sequential compactness of X to get a member of $\mathcal{S}$ which is in fact contained in some member of $\mathcal{U}$, thus contradicting the definition of $\mathcal{S}$.

There is one more result that we will need to show that a sequentially compact metric space is compact. First, we need a definition.

4.17. Definition. A metric space (X, d) is **totally bounded** if given any $r > 0$, there is a finite set $\{x_1, x_2, \cdots x_{n(r)}\} \subset X$ such that for $x \in X$ there is an i, $1 \leq i \leq n(r)$, such that $d(x, x_i) < r$.

A subset S of a metric space (X, d) is totally bounded if the metric space (S, d) is totally bounded.

Thus a metric space is totally bounded if none of its points is too far away from a finite set that is spread throughout the space. Of course the number of elements in this finite set depends on how far away "too far away" is. This is why we use $n(r)$ instead of just n in the definition of totally bounded: the number of points necessary so that every point in the space will be no more than r away from at least one of these points depends on r.

4.18. Exercises.
1) Every totally bounded subset of a metric space is bounded.
2) Total boundedness is equivalent to boundedness in $\mathbf{R}$ and in E^2, both with their usual topologies. [*Hint:* Show that a subset of $\mathbf{R}$ or E^2 is totally bounded if and only if its closure is totally bounded, and use what you know about compactness in these spaces.]
3) Let $X = \mathbf{R}$ with the discrete metric: $d(x, y) = 1$ if $x \neq y$, and $d(x, x) = 0$. Show that (X, d) is a bounded metric space which is not totally bounded. Thus, total boundedness is a stronger property than boundedness is.

To show that a sequentially compact metric space is compact, we need the following theorem. To prove it, show that if it is not true, then the space contains a sequence with no convergent subsequence. Do this by constructing such a sequence.

4.19. Theorem. Every sequentially compact metric space is totally bounded.

Finally, we can show that a sequentially compact metric space is compact. A good idea while proving it is to go through the proof using a Lebesgue number for an arbitrary open cover, and then once you see how the proof goes, go back and revise the measures of distance used (such as the Lebesgue number) to make things come out right in the end. You have probably used this technique before in ϵ-δ proofs.

4.20. Theorem. A sequentially compact metric space is compact.

Combining Theorems 4.20 and 4.12, we have the following.

4.21. Theorem. A subset of a metric space is compact if and only if it is sequentially compact.

4.22. Exercises.

Recall that a topological space is 1st countable if there is a countable local basis at each point in the space, and that every metric space is 1st countable.

1) Show that in a 1st countable space, compactness implies sequential compactness.

2) (This exercise requires a knowledge of ordinal numbers.) It is interesting to know that even though having a countable local basis at each point allows the construction of sequences that converge to the point, 1st countability alone is not enough to guarantee that sequential compactness implies compactness. Show that the space of countable ordinals, $[0, \Omega)$, is a 1st countable sequentially compact space which is not compact.

Our compactness chart now looks like:

In a general topological space,

$$\text{compact} \Rightarrow \text{countably compact} \Rightarrow \text{pseudocompact}$$
$$\nearrow$$
$$\text{sequentially compact}$$

In metric spaces,

$$\text{compact} \Rightarrow \text{countably compact} \Rightarrow \text{pseudocompact}$$
$$\Downarrow \qquad \searrow$$
$$\text{closed and} \qquad \text{sequentially compact}$$
$$\text{bounded}$$

It is true that in metric spaces, both countable compactness and pseudocompactness imply compactness (so that in metric spaces, as in **R**, the four concepts: compact, countably compact, pseudocompact and sequentially compact, are equivalent). However, we choose to postpone the proofs of these two facts until we have more information to work with.

Another variation of compactness which is somewhat different in nature than those that we have seen up to now is the following.

4.23. Definition. A topological space is **locally compact** if each point in the space has a compact neighborhood.

Because of our convention that a neighborhood of a point need not be an open set, our definition of local compactness is different from that seen in some other books. That ours is equivalent to theirs is the substance of the following theorem.

4.24. Theorem. A topological space is locally compact if and only if every point of the space is contained in an open set whose closure (relative to the whole space) is compact.

Recall that a connected space need not be locally connected. This kind of behavior does not happen with compactness and local compactness, as you can show in the following theorem.

4.25. Theorem. A compact space must also be locally compact.

4.26. Exercise.
Give an example of a locally compact space which is not compact.

Local compactness is an important concept in analysis where one does much work with the real line and the Euclidean plane, both with their usual topologies. Neither of these spaces is compact, as we have seen, but both are locally compact: you can show that **R** is locally compact quite easily with the information that we have now, but showing that the plane is locally compact is more difficult. We can show it by showing that each point in the plane is contained in a countably compact set, and then show that in the plane, countable compactness and compactness are equivalent (as they are in **R**).

4.27. Theorem. Every point in the Euclidean plane with its usual topology is contained in a neighborhood which is countably compact.
[*Hint for proof:* Show that the closed set in the plane bounded by the unit square,

$$S = \{(x, y) \in E^2 : 0 \leq x \leq 1, 0 \leq y \leq 1\}$$

is countably compact by using an argument similar to the one given for the Bolzano-Weierstrass theorem.]

4.28. Theorem. In the Euclidean plane with its usual topology, a set is compact if and only if it is countably compact.
[*Hint for proof:* For the "if" part, use the fact that the collection $\{S_{1/n}(x, y): x, y \in \mathbf{Q}\}$ is a basis for the usual topology on E^2 to show that any open cover of a subset of the plane contains a *countable* subcover. Then use the fact that countable compactness of a subset of the plane implies that every

countable open cover has a finite subcover (a fact that you should prove) to reduce this countable subcover to a finite subcover.]

Combining Theorems 4.27 and 4.28, we have the following.

4.29. Theorem. The Euclidean plane with its usual topology is locally compact.

Some of the ideas in the proof above that compactness and countable compactness are equivalent in the plane with its usual topology deserve some comment. First of all, we showed that a *countable* open cover of a countably compact subset of the plane has a finite subcover. While this is not true in a general topological space, it is true in a large class of spaces called **T_1-spaces,** and, incidentally, it certainly justifies the name "countably compact." (T_1-spaces were discussed briefly in Exercises 2.6 of Chapter 4. We will discuss them further in the next chapter.)

4.30. Exercise.
Let $X = \mathbf{Z}^+$ with topology generated by the basis

$$\{\emptyset, [n, n + 1]:n \in \mathbf{Z}^+, n \text{ odd}\}.$$

Show that this space is countably compact according to our definition, but that it has a countable open cover with no finite subcover. (If you did the exercises about T_1-spaces, show that this space is not T_1.)

Also, in the proof of Theorem 4.28, we showed that an arbitrary open cover of a subset of the plane can be reduced to a countable subcover. Spaces in which such a reduction of an arbitrary open cover to a countable subcover can be made are important ones.

4.31. Definition. A topological space is called a **Lindelöf space*** if every open cover of the space has a countable subcover.

4.32. Exercise.
Show that every compact space is Lindelöf, but there are Lindelöf spaces which are not compact. [*Hint:* Recall what we did above with the plane. Is **R** with its usual topology a Lindelöf space?]

Closely related to Lindelöf spaces are spaces which satisfy the following.

4.33. Definition. A topological space is said to be **2nd countable** (or is said to satisfy the **second axiom of countability**) if the topology on the space can be generated by a countable basis.

4.34. Exercises.

1) Every 2nd countable space is 1st countable, but there are 1st countable spaces that are not 2nd countable.

---

* After Ernst Leonard Lindelöf (1870–1946).

2) (Lindelöf's theorem.) Every 2nd countable space is a Lindelöf space. [*Hint:* If $\{U_\lambda : \lambda \in \Lambda\}$ is an open cover of a 2nd countable space X and $\{V_n : n \in \mathbf{Z}^+\}$ is a countable basis for the topology on X, then each U_λ is a union of V_n's. Consider the collection of sets obtained by choosing one U_λ for each V_n in the set $\{V_n$:there is a $\lambda \in \Lambda$ with $V_n \subset U_\lambda\}$ such that $V_n \subset U_\lambda$.]

3) The real line and the Euclidean plane, both with their usual topologies, are 2nd countable spaces, and therefore are Lindelöf spaces.

Closely related to 2nd countability is a property called *separability*, which is defined below. Recall that a subset A of a topological space X is *dense* in X if $\bar{A} = X$.

4.35. Definition. A topological space is said to be **separable** if it has a countable dense subset (i.e., a space X is separable if X contains a countable subset A such that $\bar{A} = X$).

4.36. Exercises.

1) The real line and the Euclidean plane, both with their usual topologies, are separable.

2) Every 2nd countable space is separable.

3) The converse to Problem 2 above is false: show that the Sorgenfrey line is separable but not 2nd countable.

4) Let $X = \mathbf{R}$ with the finite complement topology. Show that X is separable and Lindelöf, but is not 2nd countable. [*Hint:* A 2nd countable space must be 1st countable; a compact space must be Lindelöf.]

5) Let $X = \mathbf{R}$ with topology defined by: $U \subset X$ is open if and only if $U = \emptyset$ or $0 \in U$. Show that X is separable but not Lindelöf.

6) A subspace of a separable space need not be separable. Consider the product of two Sorgenfrey lines, and the subspace which is the diagonal $y = -x$.

7) A closed subspace of a Lindelöf space is Lindelöf, but the product of Lindelöf spaces need not be Lindelöf. [*Hint:* See Problem 6.]

8) A separable, 1st countable space is 2nd countable.

We now have the following:

$$\text{compact} \quad\quad 2\text{nd countable} \Rightarrow 1\text{st countable}$$
$$\searrow \quad \swarrow \quad\quad \searrow$$
$$\text{Lindelöf} \quad\quad\quad \text{separable}$$

and have seen that none of these implications is reversible.

5. UNIFORM CONTINUITY

We know that a function f from a metric space (X, d_1) to a metric space (Y, d_2) is continuous if for each point $x_0 \in X$, given $\epsilon > 0$ there is a $\delta > 0$ such that for $x \in X$, if $0 < d_1(x, x_0) < \delta$ then $d_2(f(x), f(x_0)) < \epsilon$. We will say that a number δ such that $0 < d_1(x, x_0) < \delta$ implies that $d_2(f(x), f(x_0)) < \epsilon$ *works* for ϵ and x_0. Thus $f: (X, d_1) \rightarrow (Y, d_2)$ is continuous if for $x_0 \in X$ and $\epsilon > 0$, there is a $\delta > 0$ that works for ϵ and x_0.

A little thought about the definition of continuity given above will convince you that the number δ seems to depend on two things. It is usually the case that the smaller you make ϵ (don't forget that you control ϵ) the smaller the corresponding δ has to be. And it can also happen that if you pick a different point $x_1 \neq x_0$ in X, then a δ that works for a given ϵ and x_0 may not be small enough to work for the same ϵ and x_1. Consider the following example.

5.1. Exercise.
1) Let $f(x) = 1/x$ for $x \in \mathbf{R}^+ = \{r \in \mathbf{R} : r > 0\}$.
   a) Show that $\delta = \frac{1}{3}$ works for $x_0 = 1$ and $\epsilon = \frac{1}{2}$.
   b) Show that $\delta = \frac{1}{3}$ does not work for $x_0 = 1$ and $\epsilon = \frac{1}{10}$.
   c) Show that $\delta = \frac{1}{3}$ does not work for $x_0 = \frac{1}{10}$ and $\epsilon = \frac{1}{2}$.
2) Let $f(x) = 1/x$ for $x \in \mathbf{R}^+$. Show that given $\epsilon > 0$, it is impossible to find a single number $\delta > 0$ that works for all $x_0 \in \mathbf{R}^+$. (Draw a graph of f and look at some x_0's on the graph to get an idea of what to do here.)

Thus, in general, the number δ can depend on both ϵ and x_0. When you think about it, we *want* δ to depend on ϵ. Indeed, the whole idea of continuity at a point x_0 is that $f(x)$ can be made close to $f(x_0)$ simply by making x close to x_0. How close we want $f(x)$ to be to $f(x_0)$ (within ϵ) determines how close x must be to x_0 (within δ). Thus δ *should* depend on ϵ. But there is really no reason why we would want δ to depend on the particular point x_0 that we are looking at. Sometimes it does, and there is nothing that we can do about it, as Exercise 5.1 (2) above shows. But sometimes it does not.

5.2. Definition. A function f from a metric space (X, d_1) to a metric space (Y, d_2) is **uniformly continuous** on X if given $\epsilon > 0$ there is a $\delta > 0$ such that for $x_1, x_2 \in X$, if $0 < d_1(x_1, x_2) < \delta$, then $d_2(f(x_1), f(x_2)) < \epsilon$.

Thus a function between two metric spaces is uniformly continuous if given $\epsilon > 0$ there is a $\delta > 0$ such that any two points in X within δ of each other are sent by f to points within ϵ of each other.

We might say (rather imprecisely) that a real-valued function of a real variable does not "tear its domain," and is uniformly continuous if its graph does not "change slope too much too fast." These imprecise ideas have to be used with care, as we will see. First, though, use the idea of a Lebesgue number to prove the following.

5.3. Theorem. Let $f:(X, d_1) \rightarrow (Y, d_2)$ be continuous. If X is compact, then f is uniformly continuous.

5.4. Exercises.

1) Show that $f:[0, 1] \rightarrow \mathbf{R}$ given by $f(x) = \sqrt{x}$ is uniformly continuous. What is the slope of the graph of f at $x = 0$? What about a uniformly continuous function not "changing slope too much or too fast"?

2) Can a function from one metric space to another be uniformly continuous if its domain is not compact?

chapter eight
Separation Axioms

Recall that in a general topological space, compact sets need not be closed. In fact, in some spaces, a set consisting of a single point need not be closed, and such singleton sets, which are in a sense the most compact sets of all, certainly ought to be closed from an intuitive point of view, since they do not have an "inside." In general, however, they are not. It turns out, though, that in the topological spaces which are of interest in applications of topology, singleton sets and in fact all compact sets are closed.

In this chapter we investigate a hierarchy of properties, called **separation axioms** most of which, when imposed on a space, will insure that compact sets are closed. For completeness, as well as historical reasons, we first look briefly at some classes of weak spaces, spaces in which compact sets need not be closed.

1. WEAK SPACES

We considered the following two classes of topological spaces in Exercises 2.6 of Chapter 4.

1.1. Definition.
1) A topological space X is a T_0-**space** if given $x, y \in X$ with $x \neq y$, either there is an open set U such that $x \in U$ and $y \notin U$, or there is an open set V such that $y \in V$ and $x \notin V$.
2) A topological space X is a T_1-**space** if given $x, y \in X$, there is an open set U such that $x \in U$ and $y \notin U$, and there is an open set V such that $y \in V$ and $x \notin V$.

Thus a topological space is a T_0-space (or, just T_0) if given any two points in the space, there is a neighborhood of at least one of them that does not contain the other, and a topological space is a T_1-space (or, just T_1) if given any two points in the space, there are neighborhoods of each of them that do not contain the other.

We say that a T_0-space is *weaker* than a T_1-space or that a T_1-space is *stronger* than a T_0-space because every T_1-space is also T_0 but there are T_0-spaces that are not T_1. You can justify this in the following exercises.

1.2. Exercises.
1) Give an example of a space that is not a T_0-space. (Thus there are spaces that are weaker than T_0.)
2) Give an example of a space that is T_0 but not T_1.
3) Show that every T_1-space is also T_0.
4) Show that every metric space is T_1 (and therefore is T_0). In particular, **R** with its usual topology is T_1.
5) State precisely what it means when a space is not a T_1-space.

T_1-spaces have some important properties that weaker spaces do not share. Prove the following theorem.

1.3. Theorem.
1) A topological space is a T_1-space if and only if for each point x in the space, $\{x\}$ is a closed set.
2) In a T_1-space, a point p is a cluster point of a set A if and only if every neighborhood of p contains infinitely many points of A.

Thus, according to the exercises above, a space is a T_1-space if and only if its points are closed sets, and in a T_1-space, a set really has to "cluster" at a cluster point because every neighborhood of a cluster point of a set must contain infinitely many points of the set.

There is an interesting equivalence of the definition of countable compactness that is valid in T_1-spaces, a property that makes it very clear why the name "countably" compact is used, and also makes it very obvious that in T_1-spaces, compactness implies countable compactness.

1.4. Theorem. A subset A of a T_1-space is countably compact if and only if every countable open cover of A has a finite subcover.

That being a T_1-space is not enough to make a space behave as well as we would like can be seen in the following exercises.

1.5. Exercises.
1) A compact subset of a T_1-space need not be closed.
2) In a T_1-space, a sequence can converge to more than one point.

3) In Theorem 1.4, is the condition that the space be T_1 needed in both directions?

2. HAUSDORFF SPACES*

The first two problems in Exercises 1.5 show that being a T_1-space is not enough to ensure that compact sets are closed or that a convergent sequence must converge to exactly one point. To get these important and useful properties, we must put additional restrictions on a space.

2.1. Definition. A topological space X is a **Hausdorff** space (also called a T_2-**space**) if given x, $y \in X$ with $x \neq y$, there are open sets U and V in X such that $x \in U$, $y \in V$ and $U \cap V = \emptyset$.

Thus a space is a Hausdorff space if and only if distinct points are contained in disjoint neighborhoods.

2.2. Exercises.
1) Show that every Hausdorff space is a T_1-space.
2) Give an example of a T_1-space which is not a Hausdorff space.
3) Show that every metric space is Hausdorff. In particular, then, **R** with its usual topology is a Hausdorff space.
4) State precisely what it means when a topological space is not a Hausdorff space.

Thus the Hausdorff property is stronger than the T_1-property; stated another way, the T_1-property is weaker than the Hausdorff property.

Since every Hausdorff space is also T_1, Hausdorff spaces enjoy all of the properties of T_1-spaces. In particular,

2.3. Theorem. In a Hausdorff space,
1) Each singleton set is closed.
2) A point p is a cluster point of a set A if and only if every neighborhood of p contains infinitely many points of A.
3) A subset A is countably compact if and only if every countable open cover of A has a finite subcover.

Thus in Hausdorff spaces, as in T_1-spaces, points are closed, a set really has to cluster at a cluster point, and countable compactness lives up to its name. But Hausdorff spaces have some nice properties that T_1-spaces need not have. Specifically,

2.4. Theorem.
1) A compact set in a Hausdorff space is closed.
2) In a Hausdorff space, a sequence can converge to at most one point.

* After Felix Hausdorff (1868–1942), one of the founders of the study of topology as such.

Thus, in a Hausdorff space, not only are singleton sets (which are always compact) necessarily closed, but *any* compact set must be closed. It is interesting to observe that the Hausdorff property on a space makes compact sets behave like singleton sets in the following way.

2.5. Theorem. If A and B are disjoint compact subsets of a Hausdorff space X, then there are disjoint open sets U and V in X such that $A \subset U$ and $B \subset V$.

Part 2 of Theorem 2.4 says that convergence of sequences is unique in Hausdorff spaces. It is interesting to know that uniqueness of convergence does not characterize the Hausdorff property of a space; in other words, there are non-Hausdorff spaces in which a sequence can converge to at most one point, as you can show in the following exercise.

2.6. Exercise.

Let $X = \mathbf{R}$ with the *countable complement* topology defined as follows: $U \subset \mathbf{R}$ is open if and only if $U = \emptyset$ or $X - U$ is countable. Show that this space is not Hausdorff but that if a sequence in this space converges, then it converges to exactly one point.

Another important property that Hausdorff spaces have is the following.

2.7. Theorem. Let X and Y be topological spaces and let $f: X \to Y$ be a continuous 1-1 function of X into Y. If X is compact and Y is Hausdorff, then f is a homeomorphism of X into Y. In particular, if f is also onto Y, then when X is compact and Y is Hausdorff, X and Y are homeomorphic.

It is important to see that Theorem 2.7 really does say something, and that the conditions on the spaces X and Y are necessary. You can do so in the following exercises.

2.8. Exercises.

1) Give an example of a compact space X and a space Y such that there is a continuous, 1-1 and onto function from X to Y, and yet X and Y are not homeomorphic. [*Hint:* A useful trick in such problems is to let X and Y be the same point-set with different topologies, and then to use the identity function.]

2) Give an example of a space X and a Hausdorff space Y such that there is a continuous, 1-1 and onto function from X to Y, and yet X and Y are not homeomorphic. [*Hint:* See Problem 1 above.]

The following properties should also be observed.

2.9. Exercises.

1) The product of Hausdorff spaces is a Hausdorff space.

2) Any subspace of a Hausdorff space is also a Hausdorff space.
3) The continuous image of a Hausdorff space need not be a Hausdorff space.
4) Any space homeomorphic to a Hausdorff space is also a Hausdorff space.

The Hausdorff property can be characterized using product spaces as follows.

2.10. Theorem. A topoligical space X is Hausdorff if and only if the diagonal $\{(x, x) : x \in X\}$ is closed in the product space $X \times X$.

This idea can also be applied to continuous functions.

2.11. Theorem. Let X and Y be topological spaces and let $f : X \to Y$ be continuous. Then the set $\{(x_1, x_2) : f(x_1) = f(x_2)\}$ is closed in $X \times Y$ if Y is a Hausdorff space.
[*Hint for proof:* Consider the function $f \times f : X \times X \to Y \times Y$.]

Retracts of Hausdorff spaces behave very well, as you can show in the following theorem.

2.12. Theorem. A retract of a Hausdorff space is closed.

2.13. Exercise.
Give an example to show that the Hausdorff property is necessary in Theorem 2.12.

3. REGULAR AND COMPLETELY REGULAR SPACES

In a Hausdorff space, distinct points are contained in disjoint open sets, so we say that distinct points can be *separated* (by disjoint open sets). A natural extension of this is to see what happens when a point can be separated from any closed set that does not contain it, rather than just from another point. You might notice that the idea of separating points from closed sets is not an extension of separating points from points unless the points of a space happen to be closed sets themselves. To avoid what is really an unnecessary complication, we will assume from now on that all spaces discussed are at least T_1-spaces (so that points are closed). In fact, we will assume more:

Unless specifically mentioned to the contrary, all topological spaces discussed in the remainder of this book will be Hausdorff spaces.

This assumption is really not unrealistic because, as mentioned earlier, all of the important topological spaces are indeed Hausdorff spaces.

With the assumption that all spaces are Hausdorff, we can generalize the Hausdorff property as follows.

3.1. Definition. A topological space X is **regular*** if whenever F is a closed subset of X and $x \in X - F$, then there are open sets U and V in X such that $x \in U, F \subset V$ and $U \cap V = \emptyset$.

Thus a space is regular if disjoint points and closed sets can be separated by disjoint open sets.

3.2. Exercises.
1) State precisely what it means when a topological space is not regular.
2) Because of our assumption that all spaces are Hausdorff, every regular space is also a Hausdorff space. But the regularity property is stronger than the Hausdorff property. To see an example of a Hausdorff space that is not regular, let X be the closed upper half plane, $X = \{(x, y) : x, y \in \mathbf{R}$ and $y \geq 0\}$, and let X have the topology generated by the following basis: for $(x, y) \in X$ with $y = 0$ (i.e., (x, y) is on the x-axis), basic open neighborhoods have the form

$$S_r(x, y) \cap \{(x, y) : y > 0\} \cup (x, 0), \qquad \text{where} \quad r > 0,$$

and for $(x, y) \in X$ with $y > 0$ (i.e., (x, y) is in the upper half plane but not on the x-axis), basic open neighborhoods of (x, y) have the form $S_r(x, y)$ where $0 < r < y$. The following picture should make this basis clear.

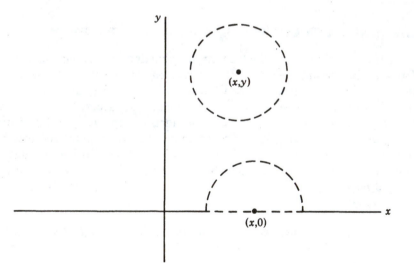

---

* In some books, a regular space is not assumed to be Hausdorff or even T_1, so you should be careful when reading other sources. The symbol T_3 is often used in connection with regular spaces.

a) Prove that the basis described above is in fact a basis for a topology on X.
b) Prove that the space described above is a Hausdorff space.
c) What is the relative topology on the x-axis inherited from the topology on X?
d) Prove that the space described above is not regular.

3) Show that every metric space is regular. In particular, then, **R** with its usual topology is regular. [Be careful! A closed set need not be bounded and need not be connected, so, in particular, a closed subset of **R** with its usual topology need not be a closed interval. Show that if F is a closed subset of a metric space (X, d) and $x \in X - F$, then there is an r-ball centered at x whose *closure* is disjoint from F.]

4) Any subspace of a regular space is regular.

5) The product of regular spaces is a regular space.

An important equivalence of regularity is the following. (We used a special case of this to show that a metric space is regular in Problem 2 above.)

3.3. Theorem. A (Hausdorff) space X is regular if and only if whenever F is closed in X and $x \in X - F$, then there is an open set U in X such that $x \in U$ and $\bar{U} \cap F = \emptyset$.

Thus a space is regular if and only if given a point and a closed set not containing the point, there is a neighborhood of the point whose closure does not meet the closed set.

We have seen that a Hausdorff space need not be regular. However, we can draw on some of what we have discussed previously and show that if a Hausdorff space is locally compact, then it is also regular. To do this, we first introduce a new idea, which is a generalization of adding the "point at infinity" to a plane to get a sphere.

3.4. Definition. Let X be a (Hausdorff) space which is not compact, and let p be an object that is not a member of X. The Alexandroff* **one-point compactification** of X will be denoted here by $\tilde{X}$, and is defined to be $X \cup \{p\}$, with topology defined by

1) neighborhoods of points of X are the same as in the topology on X, and
2) $U \subset \tilde{X}$ is a basic open neighborhood of p if and only if $p \in U$ and $\tilde{X} - U$ is a compact subset of X.

3.5. Theorem. The one-point compactification of a Hausdorff space X is a compact space which contains X as a dense subspace.

* After P. S. Alexandroff (1896–   ).

We have said that all of our spaces will be Hausdorff spaces, so we had better be careful when building new spaces (like the one-point compactification) to make sure that these new spaces are Hausdorff. As a matter of fact, without an additional restriction on the space that we start with, it turns out that the one-point compactification of a Hausdorff space need *not* be Hausdorff! But this restriction is very natural, though, because it also turns out that if X is Hausdorff, then X must automatically satisfy this additional restriction. The restriction is simply that X, in addition to being Hausdorff, must also be locally compact.

3.6. Theorem. The one-point compactification of a Hausdorff space X is a Hausdorff space if and only if X is locally compact.

Using the one-point compactification and Theorem 2.5, you can show the following.

3.7. Theorem. A locally compact Hausdorff space is regular.

In Hausdorff spaces we can separate distinct points by disjoint open sets, and in regular spaces, we can separate a point from a closed set that does not contain the point by disjoint open sets. The next logical step in this hierarchy of stronger conditions on a (Hausdorff) space would be to impose the condition on a space that disjoint closed sets can be separated by disjoint open sets. Before discussing such a condition, we digress for a moment and look at spaces in which a point that does not belong to a closed set can be separated from the closed set in a special way. Let I denote the closed unit interval $[0, 1]$ with its usual topology.

3.8. Definition. A (Hausdorff) space X is **completely regular*** if whenever F is a closed subset of X and $x \in X - F$, then there is a continuous function $f : X \to I$ such that $f(x) = 0$ and $f(F) = 1$.

We say that a closed subset F of a topological space X and a point $x \in X - F$ can be *separated by a continuous function* if there is a continuous function $f : X \to I$ such that $f(x) = 0$ and $f(F) = 1$. Thus a space is completely regular if and only if disjoint points and closed sets can be separated by continuous functions.

The following theorem should be easy. If it is not, go back and read Theorem 2.8 in Chapter 6 again.

3.9. Theorem. Every completely regular space is regular.

Completely regular spaces are important because when a non-trivial Hausdorff space is completely regular, we can be sure that there are non-

---

* See the note on regular spaces. Completely regular spaces are sometimes called $T_{3 1/2}$-spaces, and are also called Tychonoff spaces (after A. N. Tychonoff).

constant real-valued continuous functions defined on it (Why can we?). Surprisingly, there are regular spaces such that every real-valued continuous function defined on the space is constant. (One such space is a variation of a space called the *Tychonoff corkscrew*, and is due to E. Hewitt. This space is quite complicated and requires an understanding of ordinal numbers to construct. You can find a good description of it in Steen and Seebach [17].)

Of course not every regular but not completely regular space has the property that every real-valued continuous function defined on it is constant, as we will see. The difference between regularity and complete regularity is that in a regular space, *some* disjoint points and closed sets may be separated by continuous functions and *all* disjoint points and closed sets can be separated by disjoint open sets, while in a completely regular space, *all* disjoint points and closed sets can be separated by continuous functions (and therefore, all disjoint points and closed sets can be separated by disjoint open sets). Separation by continuous functions is a stronger property than separation by disjoint open sets.

3.10. Exercises.

1) The real line with its usual topology is completely regular. [*Hint:* To prove this, you must construct a continuous function from **R** into *I* which separates an arbitrary (but fixed) closed subset of **R** from an arbitrary (but fixed) point of **R** not in the closed set. It might be easier to draw a picture of **R** with the closed set and the point on the picture, draw a graph of the function that you want, and then define the function by looking at its graph.]

2) Any discrete space is completely regular.

3) A subspace of a completely regular space is completely regular.

4) Show that a topological space X is completely regular if and only if whenever F is a closed subset of X and $x \in X - F$, then there is a continuous function $f: X \rightarrow I$ such that $f(x) = 1$ and $f(F) = 0$. More generally, show that X is completely regular if and only if there is a continuous function $f: X \rightarrow [a, b]$, where $a, b \in \mathbf{R}$ with $a < b$, and $[a, b]$ has its usual topology, such that $f(x) = a$ and $f(F) = b$.

5) State precisely what it means when a topological space is not completely regular.

6) The continuous image of a completely regular space need not be completely regular.

7) The homeomorphic image of a completely regular space is completely regular.

The rest of this section requires a knowledge of infinite products.

Completely regular spaces satisfy an interesting and important property: as we will show, every completely regular space is a topological subset of a product space whose factors are copies of the unit interval $I = [0, 1]$ with its usual topology; furthermore, the "number" of factors in this product space is the same as the "number" of continuous functions from the space into I. In other words, every completely regular space is homeomorphic to a subset of a product of unit intervals. Before proving this, we establish some facts about product spaces indexed by functions.

Let $\mathfrak{F}$ denote a set of continuous functions from a space X into I. For each $f \in \mathfrak{F}$, let $I_f = I$ and let $P = \prod_{f \in \mathfrak{F}} I_f$ have the product topology. We define a function from X into P as follows: define $\varphi : X \to P$ by $\varphi(x) = (f(x))_{f \in \mathfrak{F}}$. This deserves some comment. We want φ to be a function from X to the product space P, so the value of φ at a point $x \in X$ is a point in this product. Now points in a product consist of coordinates, one from each factor, so we can completely determine the point $\varphi(x)$ if we say what its coordinates are. In our situation, the factors of P are copies of I, and there is one such factor for each $f \in \mathfrak{F}$. Hence we only need to specify the "f-th coordinate" of $\varphi(x)$ for an arbitrary $f \in \mathfrak{F}$ in order to completely determine $\varphi(x)$. The obvious way to do this is to let the "f-th coordinate" of $\varphi(x)$ be the point $f(x)$ in $I_f = I$. Then $\varphi(x)$ is the point $(f(x))_{f \in \mathfrak{F}}$ in the product space P.

3.11. Exercises.
Let P and φ be defined as above.
1) For $f \in \mathfrak{F}$, let $\pi_f : P \to I_f$ be the projection function from P into I_f. What is $\pi_f(x)$ for $x \in X$?
2) For $f \in \mathfrak{F}$, what is $\pi_f \circ \varphi$? Is it continuous? Is φ continuous?

We are interested in having the function $\varphi : X \to P$ be a homeomorphism of X into P, which means that we want φ to be continuous and 1-1, and we also want $\varphi^{-1} : \varphi(X) \to X$ to be continuous. We have seen (Problem 2 above) that φ is continuous when $\mathfrak{F}$ is a collection of continuous functions of X into I. To see that the other conditions are not automatically satisfied, consider the following (rather trivial) example.

3.12. Exercises.
Let $X = \mathbf{R}$ with the indiscrete topology and let $\mathfrak{F} = C(X, I)$, the set of all continuous functions from X into I. For each $f \in \mathfrak{F}$, let $I_f = I$ and define $P = \prod_{f \in \mathfrak{F}} I_f$, as above.
1) Describe the set $\mathfrak{F}$. In other words, what does a continuous function from X into I look like?
2) Define $\varphi : X \to P$ as above. What is $\varphi(X)$? Is φ 1-1?

The problem with the example above is that there are not enough continuous functions from X into I. When there *are* enough continuous functions from X into I, the function φ behaves much better. You might expect that complete regularity will enter into this discussion, because complete regularity of a space X is defined in terms of continuous functions from X into I, and a completely regular space has "lots" of continuous functions defined on it with range in I(enough to separate all disjoint points and closed sets). As we will see, complete regularity of X makes φ behave very well indeed.

We say that a collection of functions $\mathfrak{F}$ from X into I **separates points** if for x, $y \in X$ with $x \neq y$, there is a function $f \in \mathfrak{F}$ such that $f(x) \neq f(y)$. Similarly, we say that a collection of functions $\mathfrak{F}$ from X to I **separates points and closed sets** if whenever F is a closed subset of X and $x \in X - F$, then there is a function $f \in \mathfrak{F}$ such that $f(x) \notin \overline{f(F)}$.

3.13. Theorem. Let $\mathfrak{F}$ be a collection of continuous functions from a space X into I. For each $f \in \mathfrak{F}$, let $I_f = I$, and let $P = \prod_{f \in \mathfrak{F}} I_f$ with the product topology. Define $\varphi: X \to P$ by $\varphi(x) = (f(x))_{f \in \mathfrak{F}}$. Then φ is continuous, and if $\mathfrak{F}$ separates points, φ is 1-1. If, in addition, $\mathfrak{F}$ separates points and closed sets, $\varphi^{-1}: \varphi(X) \to X$ is continuous.

[*Hint for proof:* Show that to prove that φ^{-1} is continuous, it is sufficient to show that φ itself is an open map, i.e., that $\varphi(U)$ is open whenever U is open.]

We can now prove the property of completely regular spaces mentioned earlier.

3.14. Theorem. Every completely regular space X is homeomorphic to a subspace of a product of unit intervals.

[*Hint for proof:* Consider $C(X, I)$, the collection of all continuous functions from X into I.]

To show that the product of completely regular spaces is completely regular, we first establish the fact that a space is completely regular if and only if a point can be separated from the complement of a *sub-basic* open neighborhood of the point by a continuous function.

3.15. Theorem. Let $\mathcal{S}$ be a sub-basis for the topology on a (Hausdorff) space X. Then X is completely regular if and only if whenever $x \in X$ and $S \in \mathcal{S}$ with $x \in S$, then there is a continuous function $f: X \to I$ such that $f(x) = 0$ and $f(X - S) = 1$.

3.16. Theorem. The product of completely regular spaces is completely regular.

4. NORMAL SPACES

In a Hausdorff space, distinct points can be separated by disjoint open sets; in a regular space, disjoint points and closed sets can be separated by disjoint open sets; In a completely regular space, disjoint points and closed sets can be separated by continuous functions (and therefore can be separated by disjoint open sets). We have seen that complete regularity implies regularity, and regularity implies Hausdorff (this last implication because of our assumption that all spaces are Hausdorff).

There are two ways to strengthen the separation properties required of a space at this point: we could generalize regularity and ask that two disjoint closed subsets of a space be separated by disjoint open sets, or we could generalize complete regularity and ask that two disjoint closed subsets of a space be separated by continuous functions (where F_1 and F_2 are separated by $f: X \rightarrow I$ if $f(F_1) = 0$ and $f(F_2) = 1$). That these two approaches to strengthening the separation properties of a space are in fact the same is the subject of a famous and difficult theorem due to P. Urysohn. This theorem says that in a Hausdorff space, two disjoint closed sets can be separated by disjoint open sets if and only if they can be separated by a continuous function. Spaces with these properties are called **normal** spaces.* As we will see (and as Kelley [9] observes), normal spaces are distinctly not "normal" in their behavior: for example, subspaces of normal spaces need not be normal and products of normal spaces need not be normal. Before we can show any of this, we need a definition

4.1. Definition. A (Hausdorff) topological space X is **normal** if whenever F_1 and F_2 are disjoint closed subsets of X, then there are disjoint open subsets U_1 and U_2 of X such that $F_1 \subset U_1$ and $F_2 \subset U_2$.

Thus a space is normal if disjoint closed sets can be separated by disjoint open sets. Note that if the disjoint closed sets are *compact*, then a Hausdorff space has this property. Thus normality can be thought of as a generalization of the Hausdorff property. In fact, if a Hausdorff space is compact, then it is automatically normal. This generalizes the fact that a locally compact Hausdorff space is regular.

4.2. Theorem. A compact Hausdorff space is normal.

4.3. Exercises.
1) Every normal space is regular.
2) A discrete space is normal.
3) State precisely what it means when a topological space is not normal.
4) The continuous image of a normal space need not be normal.
5) The homeomorphic image of a normal space is normal.
6) Every metric space is normal.

Urysohn's theorem says that our definition of normality is equivalent to

* See the notes on regular and completely regular spaces. The symbol T_4 is also used in connection with normality.

the fact that disjoint closed sets can be separated by continuous functions. To prove that our definition implies separation by continuous functions involves constructing a continuous function from a topological space into I which separates two given closed subsets of the space; as is usually the case with such constructive proofs, this is quite difficult. The proof of the equivalence in the other direction, however, is quite easy, and you can do it now. See the proof of Theorem 3.9.

4.4. Theorem. If for every pair of disjoint closed subsets F_1 and F_2 of a topological space X there is a continuous function $f: X \to I$ such that $f(F_1) = 0$ and $f(F_2) = 1$, then X is normal.

For the converse of Theorem 4.4, we need to construct a continuous function from X to I which separates two given disjoint closed subsets of X, when we know that any two disjoint closed subsets of X can be separated by disjoint open sets. Before doing this, we establish the following result.

4.5. Theorem. A topological space X is normal if and only if whenever F is closed in X and U is open in X with $F \subset U$, then there is an open set V in X such that $F \subset V \subset \bar{V} \subset U$.

We can now prove that normality implies that disjoint closed sets can be separated by continuous functions. This result is known as **Urysohn's lemma.**

4.6. Theorem. Let X be a normal topological space, and let F_1 and F_2 be disjoint closed subsets of X. Then there is a continuous function $f: X \to I$ such that $f(F_1) = 0$ and $f(F_2) = 1$.

Outline of proof: Show that there is a collection of open sets $\{U_r : r = m/2^n, n \in \mathbf{Z}^+, m = 1, 2, \cdots, 2^n - 1\}$ such that for $r_1 < r_2$, $F_1 \subset U_{r_1} \subset \bar{U}_{r_1} \subset U_{r_2} \subset \bar{U}_{r_2} \subset X - F_2$. Define $f: X \to I$ by $f(x) = 0$ if $x \in U_r$ for all r, and $f(x) = \sup\{r : x \notin U_r\}$ if there is an r such that $x \notin U_r$. Show that $f(F_1) = 0$, $f(F_2) = 1$, and $0 \leq f(x) \leq 1$ for all $x \in X$. The only thing left is to show that f is continuous. To do this, use the sub-basis $\mathcal{S} = \{[0, a), (b, 1] : 0 \leq a, b \leq 1\}$ for the topology on I, and show that $f^{-1}(S)$ is open in X for each $S \in \mathcal{S}$.

As an immediate result of Urysohn's lemma, we have the following.

4.7. Corollary. Every normal space is completely regular.

It is true that not every completely regular space is normal (so that normality is stronger than complete regularity). However, we will have to wait to see an example of a non-normal completely regular space until we have more information to work with.

Combining Urysohn's lemma with Theorem 4.4, we have the following characterization of normality.

4.8. Theorem. A topological space X is normal if and only if whenever F_1 and F_2 are disjoint closed subsets of X, then there is a continuous function $f: X \to I$ such that $f(F_1) = 0$ and $f(F_2) = 1$.

It is important to note the difference between normality and regularity. In a regular space, a point and a closed set that does not contain the point can be separated by disjoint open sets, but they may not be able to be separated by a continuous function; spaces in which disjoint points and closed sets can be separated by continuous functions (completely regular spaces) are more than just regular. In a normal space however, the two separation methods—separation of disjoint closed sets by disjoint open sets, and separation of disjoint closed sets by continuous functions—are equivalent.

The following corollary to Urysohn's theorem (Theorem 4.8) is often useful.

4.9. Theorem. A topological space is normal if and only if whenever F_1 and F_2 are disjoint closed subsets of X and $a, b \in \mathbf{R}$ with $a < b$, then there is a continuous function $f: X \to [a, b]$ (where $[a, b]$ has its usual topology), such that $f(F_1) = a$ and $f(F_2) = b$.

Our next result concerning normal spaces is due to H. Tietze, and says that a space is normal if and only if a bounded continuous real-valued function on a closed subset of the space can be extended to a continuous function on the whole space.* We discussed extensions of functions briefly (and in a rather specialized setting) when we proved that pseudocompactness is equivalent to compactness in $\mathbf{R}$. Recall that if X and Y are topological spaces and if $A \subset X$, then a continuous function $f: A \to Y$ can be *extended* over X if there is a continuous function $F: X \to Y$ such that $F \mid A = f$. The function F is a *continuous extension* of f.

As with Urysohn's theorem, half of Tietze's theorem is not difficult, and we can prove it now.

4.10. Theorem. Let X be a (Hausdorff) topological space such that whenever C is a closed subset of X and $f: C \to I$ is continuous, there is a continuous extension of f to X. Then X is normal.

[*Hint for proof:* Let A and B be closed disjoint subsets of X and put $C = A \cup B$. Extend an appropriate $f: C \to \mathbf{R}$ to see that X is normal.]

For the other half of Tietze's theorem, we must construct a continuous extension of a continuous function on a closed subset of a normal space. Before doing this, we first establish a result about continuous functions, which is a generalization of the Weierstrass M-test from analysis.

4.11. Theorem. Let $\{f_n : n \in \mathbf{Z}^+\}$ be a sequence of continuous functions

---

* The requirement that the function must be bounded can be deleted in Tietze's theorem. See Theorem 4.13.

from a topological space X into $\mathbf{R}$, and let $\{M_n : n \in \mathbf{Z}^+\}$ be a sequence of real numbers such that $\sum_{n=1}^{\infty} M_n$ exists. If for each $n \in \mathbf{Z}^+$, $|f_n(x)| \leq M_n$ for all $x \in X$, then the function $f(x) = \sum_{n=1}^{\infty} f_n(x)$ is a continuous function from X into $\mathbf{R}$.

We can now prove the other half of Tietze's theorem. This result is known as the **Tietze extension theorem**.

4.12. Theorem. Let X be a normal topological space and let C be a closed subset of X. Then a continuous function $f : C \to [-1, 1]$ (with its usual topology) can be extended to a continuous function $F : X \to [-1, 1]$.

Outline of Proof: Define sequences of functions $\{f_n : n \in \mathbf{Z}^+\}$ and $\{g_n : n \in \mathbf{Z}^+\}$ as follows: put $f_1 = f$. Let $A_1 = \{x \in X : f_1(x) \leq -\frac{1}{3}\}$ and $B_1 = \{x \in X : f_1(x) \geq \frac{1}{3}\}$. Show that A_1 and B_1 are disjoint closed subsets of X, and show that there is a continuous function $g_1 : X \to [-\frac{1}{3}, \frac{1}{3}]$ such that $g_1(A_1) = -\frac{1}{3}$ and $g_1(B_1) = \frac{1}{3}$. Define $f_2 = f_1 - g_1$ and show that $|f_2(x)| \leq \frac{2}{3}$ for all $x \in X$. Now put $A_2 = \{x \in X : f_2(x) \leq -\frac{2}{9}\}$ and $B_2 = \{x \in X : f_2(x) \geq \frac{2}{9}\}$. Get $g_2 : X \to [-\frac{2}{9}, \frac{2}{9}]$ such that $g_2(A_2) = -\frac{2}{9}$ and $g_2(B_2) = \frac{2}{9}$. Put $f_3 = f_2 - g_2 = f_1 - (g_1 + g_2)$ and show that $|f_3(x)| \leq (\frac{2}{3})^2$ for all $x \in X$. Continue by induction to get $\{f_n : n \in \mathbf{Z}^+\}$ with $|f_n(x)| \leq (\frac{2}{3})^{n-1}$ for all $x \in X$, and $\{g_n : n \in \mathbf{Z}^+\}$ with $|g_n(x)| \leq \frac{1}{3}(\frac{2}{3})^{n-1}$, such that $f_n = f_1 - (g_1 + g_2 + \cdots + g_{n-1})$. Put $F(x) = \sum_{n=1}^{\infty} g_n$ and show that F is a continuous extension of f to all of X, and the range of F is contained in $[-1, 1]$.

Now that we have the Tietze extension theorem, the following useful generalization of it is not hard to prove. We combine the Tietze extension theorem with a modification of Theorem 4.10 to get the following characterization of normality.

4.13. Theorem. Let X be a topological space and let C be a closed subset of X. Let $a, b \in \mathbf{R}$ with $a < b$, and let $[a, b]$ have its usual topology. Then every continuous function $f : C \to [a, b]$ can be extended to a continuous function $F : X \to [a, b]$ if and only if X is normal.

Furthermore, if f is any continuous function from C into $\mathbf{R}$, bounded or not, then there is a continuous extension F of f to all of X.

Using the Tietze extension theorem and the fact that a metric space is normal, we can prove that pseudocompactness is equivalent to countable compactness in metric spaces. Since we already know that countable compactness always implies pseudocompactness (in any topological space— not just in metric spaces), we only need to show that pseudocompactness implies compactness in metric spaces. The proof is analogous to the proof that we gave for Theorem 3.13 in Chapter 7.

4.14. Theorem. A pseudocompact metric space is countably compact.

As we mentioned earlier, despite its nice properties, normality is distinctly not "normal" in much of its behavior. For example, as we will see, a subspace of a normal space need not be normal, and the product of normal spaces need not be normal. We will have to postpone an example of a normal space with a non-normal subspace until later, but we can see an example of a non-normal product with normal factors now. To do it without using ordinal numbers, we first need a result of F. B. Jones, which is given here as it appears in Dugundji [4].

4.15. Lemma. Let X be a topological space with a dense subset D and a closed discrete subspace S. If $|S| \geq 2^{|D|}$, then X is not normal.

Outline of Proof: Show that if X is normal, then for a subset A of S, there are disjoint open sets U_A and, V_A in X such that $A \subset U_A$ and $S - A \subset V_A$. Similarly, for $B \subset S$, there are disjoint open sets U_B and V_B in X with $B \subset U_B$ and $S - B \subset V_B$. Show that if $A \neq B$, then $D \cap U_A \neq D \cap U_B$. Finally, show that this forces the function $\varphi : \mathcal{P}(S) \to \mathcal{P}(D)$ defined by $\varphi(A) = D \cap U_A$ to be 1-1, and show that this is impossible.

4.16. Exercise.

The product of normal spaces need not be normal.
  1) (Not using ordinal numbers.) Let $X = \mathbf{R}$ with the Sorgenfrey topology.
     a) X is normal. [*Hint:* Let A and B be disjoint closed subsets of X. For each $a \in A$, let $b_a = \inf \{b \in B : b > a\}$. Show that $A \subset \bigcup_{a \in A} [a, b_a)$. Similarly, define a_b and show that $B \subset \bigcup_{b \in B} [b, a_b)$.
     b) $X \times X$ (with the product topology) is not normal. [*Hint:* Conside the subspace $\{(x, y) : y = -x \text{ and } x \text{ is irrational}\}$.)]
  2) (Using ordinal numbers.) Both $[0, \Omega)$ and $[0, \Omega]$ (with their usual order topologies) are normal. [*Hint:* See the hint to Problem 1 (a) above.]
     $[0, \Omega] \times [0, \Omega)$ (with the product topology) is not normal. [*Hint:* Consider the diagonal $\{(x, y) : x = y\}$ and the "edge" $\{\Omega\} \times [0, \Omega)$. (We will show later that $[0, \Omega] \times [0, \Omega]$ is normal, so here we have an example of a normal space with a non-normal subspace.)]

While we are on the subject of normality and product spaces, we remark that it was unknown for some time whether or not there exists a normal space whose product with the unit interval is not normal. This question, known as the *binormality problem*, was recently settled by Mary Ellen Rudin, who exhibited a (very complicated) normal space whose product with the unit interval is not normal [12].

As mentioned earlier, it is not true that every subspace of a normal space is normal, but it is true that a *closed* subspace of a normal space is normal.

4.17. Theorem. Let X be a normal space and let F be closed in X. Then when F is given the relative topology inherited from the topology on X, F is normal.

It is also true that while an arbitrary subset of a normal space need not be normal, it is completely regular.

4.18. Theorem. Let X be a normal space and let $S \subset X$. Then when S is given the relative topology inherited from the topology on X, S is completely regular.

Using Theorem 4.17 and the one-point compactification (Definition 3.4), we can show that not only is a locally compact Hausdorff space regular (Theorem 3.7), but in fact, it is completely regular.

4.19. Theorem. A locally compact Hausdorff space is completely regular.

Thus combining some ideas from our study of compactness with some ideas from our study of separation axioms, we have two important results: a locally compact Hausdorff space is completely regular (and therefore is regular), and a compact Hausdorff space is normal.

chapter nine
Complete Spaces

We have seen that a subset of **R** with its usual topology is compact if and only if it is closed and bounded, but that in a general metric space, a closed and bounded set need not be compact. (For example, $[0, 1] \cap \mathbf{Q}$ is a closed and bounded subset of **Q** with its usual metric, but is not compact.) In this chapter, we will investigate the property that makes a closed and bounded subset of **R** compact, and will get another equivalence of compactness in metric spaces.

1. DEFINITION OF COMPLETENESS AND SOME CONSEQUENCES

It should be clear that if a sequence in a metric space converges, then its terms get arbitrarily close together as $n \to \infty$. However, even if the terms of a sequence in a metric space get arbitrarily close together, the sequence may not converge. Sequences in which the terms get close together are called *Cauchy* sequences.* More precisely,

1.1. Definition. A sequence $\{x_n : n \in \mathbf{Z}^+\}$ in a metric space (X, d) is called a **Cauchy sequence** if given $\epsilon > 0$, there is an $N \in \mathbf{Z}^+$ such that if $n, m \geq N$, then $d(x_n, x_m) < \epsilon$.

The following theorem is an easy consequence of the triangle inequality.

1.2. Theorem. If a sequence in a metric space converges, then it is a Cauchy sequence.

As mentioned above, the converse to Theorem 1.2 is false, as you can show in Problem 1 of the exercises below.

---

* After Augustin Louis Cauchy (1789–1857) whose work was of great importance in putting analysis on a rigorous foundation.

1.3. Exercises.

1) Give an example of a sequence in a metric space which is a Cauchy sequence, but which does not converge.
2) Give an example of a sequence in a metric space which is not a Cauchy sequence. Can your example be a convergent sequence?
3) Give an example of two metric spaces X and Y which are homeomorphic and a sequence which is a Cauchy sequence in X but whose image under the homeomorphism is not a Cauchy sequence in Y. Can the sequence converge in either space?

As Problem 3 above shows, the property of being a Cauchy sequence is *not* a topological property—it is not necessarily preserved by a homeomorphism.

The part of a sequence consisting of all terms of the sequence after some particular term is often called a "tail" of the sequence. For a sequence $\{x_n : n \in \mathbf{Z}^+\}$, let $T(x_n) = \{x_i : i \in \mathbf{Z}^+ \text{ and } i \geq n\}$.

1.4. Theorem. Let $\{x_n : n \in \mathbf{Z}^+\}$ be a Cauchy sequence in a metric space (X, d). Then given $\epsilon > 0$, there is an $n \in \mathbf{Z}^+$ such that $\delta(T(x_n)) < \epsilon$ (where $\delta(T(x_n))$ is the diameter of the set $T(x_n)$:

$$\delta(T(x_n)) = \sup \{d(x_i, x_j) : x_i, x_j \in T(x_n)\}).$$

As we have seen, a Cauchy sequence in a metric space need not converge. When all Cauchy sequences in a metric space do converge, the space is said to be *complete*.

1.5. Definition. A metric space is **complete** if every Cauchy sequence in the space converges (to a point of the space).

1.6. Exercise.

State precisely what it means when a metric space is not complete, and give an example of a non-complete metric space.

It is a fundamental fact of analysis that both the real line and the plane are complete spaces when given their usual metrics. (Indeed, the existence of least upper bounds and greatest lower bound in $\mathbf{R}$ is a consequence of the completeness of $\mathbf{R}$, as we will show.) To prove that $\mathbf{R}$ and E^2 are complete, we prove a more general theorem which says that any locally compact metric space is complete. First we need a preliminary result.

1.7. Lemma. Let (X, d) be a metric space. If $x_0 \in X$ is an accumulation point of a Cauchy sequence in X, then the sequence converges to x_0.

1.8. Exercise.

Give an example to show that the word "Cauchy" cannot be omitted in Lemma 1.7. In other words, give an example of a sequence in a metric space which has an accumulation point but which does not converge to this accumulation point. Must a subsequence of the sequence converge to the accumulation point?

To prove that a locally compact metric space is complete, recall that every infinite subset of a compact metric space has an accumulation point.

1.9. Theorem. A locally compact metric space is complete.

1.10. Corollary. The real line and the Euclidean plane are complete metric spaces when given their usual metrics.

Since Cauchy sequences are not necessarily preserved by a homeomorphism, it should not be too surprising that completeness, unlike the other properties that we have discussed, is *not* a topological property.

1.11. Exercises.
1) Give an example of two homeomorphic spaces, one of which is complete, while the other is not.
2) Prove that if A is a non-empty subset of R which is bounded above, then sup$A$ exists. (A is bounded above if there is an $n \in Z^+$ such that $A \subset (-\infty, n]$.)
3) Prove that if A is a non-empty subset of R which is bounded below, then inf$A$ exists. (A is bounded below if there is an $n \in Z^+$ such that $A \subset [-n, \infty)$.)
4) We have said that completeness is not a topological property, and in Problem 1 above, you gave an example of two homeomorphic spaces, one of which is complete and one of which is not.

   Actually, completeness is not a property of the topology on a space at all. Instead, it is a property of the metric. Recall that the topology generated by a metric d on a set X is the topology generated by the basis $\{S_r(x) : x \in X\}$, where $S_r(x) = \{y \in X : d(x,y) < r\}$.

   Define a metric d' on R as follows:

   $$d'(x, y) = \left| \frac{x}{1 + |x|} - \frac{y}{1 + |y|} \right|.$$

   a) Show that d' is a metric on R. (This is messy, as is (b) below.)
   b) Show that the topology generated by d' is the same as the usual topology generated by the usual absolute value metric d on R.
   c) Show that (R, d') is not complete. [*Hint:* Consider Z^+ as a sequence in (R, d').]

   Thus (R, d) and (R, d') have the same topology, but (R, d) is complete, whereas (R, d') is not. Completeness is a property of the metric on a space rather than the topology.

2. COMPLETENESS AND COMPACTNESS

We repeat the definition of total boundedness here, modified to apply directly to a subset of a metric space.

2.1. Definition. A subset S of a metric space (X, d) is **totally bounded** if given $r > 0$ there is a finite set $\{x_1, x_2, \cdots, x_{n(r)}\} \subset S$ such that for any $x \in S$ there is an i, $1 \leq i \leq n(r)$, such that $d(x, x_i) < r$.

Recall (Exercises 4.18 of Chapter 7) that every totally bounded subset of a metric space is bounded, but there are metric spaces in which a bounded set need not be totally bounded. Recall also that the number of elements necessary in the finite set $\{x_1, x_2, \cdots, x_{n(r)}\}$ depends on r.

Our goal in this section is to characterize compactness in metric spaces by proving that a subset of a metric space is compact if and only if it is complete and totally bounded. Part of the work has already been done. We know that compactness is equivalent to sequential compactness in metric spaces, and that a sequentially compact subset of a metric space is totally bounded. Thus, a compact subset of a metric space is totally bounded. If we can show that compactness implies completeness in metric spaces, half of our theorem will be proved. It is actually quite easy to show that a sequentially compact subset of a metric space is complete, and you can then use the equivalence of compactness and sequential compactness in metric spaces to get the result.

2.2. Theorem. A sequentially compact subset of a metric space is complete.

2.3. Corollary. A compact subset of a metric space is complete.

Thus, half of our characterization of compactness in metric spaces is proved, and we have the following.

2.4. Theorem. A compact subset of a metric space is complete and totally bounded.

To reverse this and show that a complete and totally bounded subset of a metric space is compact, the following modification of the definition of total boundedness will be useful.

2.5. Theorem. Let (X, d) be a totally bounded metric space. For any $r > 0$, there is a finite collection of r-balls $\{S_r(x_i) : 1 \leq i \leq n(r)\}$ such that $X \subset \bigcup_{i=1}^{n} S_r(x_i)$.

To prove that a complete and totally bounded subset of a metric space is compact, we will show that such a subset is sequentially compact. The method suggested is adapted from that found in Simmons [14]. The notation used is meant to keep things clear. Don't let it confuse you. The ideas used are familiar.

2.6. Theorem. Let A be a complete and totally bounded subset of a metric space (X, d). Then A is sequentially compact.

Outline of Proof: Let $S_1 = \{x_1{}^1, x_2{}^1, x_3{}^1, \cdots\}$ be a sequence of points of A. Show that there is an $x_1 \in A$ and a subsequence $S_2 = \{x_1{}^2, x_2{}^2, x_3{}^2, \cdots\}$ of S_1 such that $S_2 \subset S_{1/2}(x_1)$. Now show that S_2 has a subsequence $S_3 = \{x_1{}^3, x_2{}^3, x_3{}^3, \cdots\}$ such that there is a point $x_2 \in A \cap \bar{S}_{1/2}(x_1)$ with $S_3 \subset S_{1/3}(x_2)$. Continue this and show that the sequence $\{x_1{}^1, x_2{}^2, x_3{}^3, \cdots\}$ obtained by taking the "diagonal" element from each S_n, is a subsequence of S_1 which is a Cauchy sequence.

2.7. Corollary. A complete and totally bounded subset of a metric space is compact.

Combining this corollary with Theorem 2.4, we have the following characterization of compactness in metric spaces.

2.8. Theorem. A subset of a metric space is compact if and only if it is totally bounded.

We can use this characterization of compactness in metric spaces to show that countable compactness is equivalent to compactness in metric spaces. Since we already know that compact always implies countably compact (in any topological space—not just in metric spaces), we need only show that countably compact implies compact in metric spaces.

2.9. Theorem. A countably compact subset of a metric space is complete and totally bounded.

2.10. Corollary. A countably compact subset of a metric space is compact.

If a metric space itself is complete, we have the following results.

2.11. Theorem. Let (X, d) be a complete metric space. Then $S \subset X$ is complete (i.e., (S, d) is a complete metric space) if and only if S is a closed subset of X.

2.12. Theorem. A closed subset of a complete metric space is compact if and only if it is totally bounded.

3. THE BAIRE CATEGORY THEOREM AND SOME APPLICATIONS

Recall that a decreasing sequence $F_1 \supset F_2 \supset F_3 \supset \cdots$ of closed subsets of a compact space has a non-empty intersection. When the compact space in question is a metric space, this property is really a consequence of the completeness of the space, as you can show in the following theorem, which is due to G. Cantor.

3.1. Theorem. Let $F_1 \supset F_2 \supset F_3 \supset \cdots$ be a decreasing sequence of closed subsets of a complete metric space. Then $\bigcap_{n=1}^{\infty} F_n \neq \emptyset$. Furthermore, if $\delta(F_n) \rightarrow 0$ as $n \rightarrow \infty$ (i.e., if $\lim_{n\to\infty} \delta(F_n) = 0$), then $\bigcap_{n=1}^{\infty} F_n$ consists of exactly one point.

Our next theorem is rather technical in nature, but it has an important and very useful corollary.

3.2. Theorem. Let $\{U_n : n \in \mathbf{Z}^+\}$ be a countable collection of open subsets of a complete metric space (X, d) such that for each n, $\bar{U}_n = X$ (i.e., each U_n is dense in X). Then $\bigcap_{n=1}^{\infty} U_n \neq \emptyset$.

[*Hint for proof:* Construct a Cauchy sequence by induction (and with some care) that converges to a point which, by the way that your sequence is constructed, must be in every U_n.]

Using Theorem 3.2 and the De Morgan laws, you can prove a corollary to Theorem 3.2 which is known as the **Baire category theorem.*** To state it in the standard way, we need several definitions.

3.3. Definition. A subset A of a topological space X (not necessarily a metric space) is said to be **nowhere dense** if $\bar{A}$ contains no non-empty open sets.

The following simple result often makes the concept of a nowhere dense set easier to use.

3.4. Theorem. A subset A of a topological space X is nowhere dense if and only if $X - \bar{A}$ is dense in X (i.e., $A \subset X$ is nowhere dense if and only if $\overline{X - \bar{A}} = X$).

3.5. Definition. A topological space is said to be of the **first category** if it is the union of a countable number of nowhere dense subsets. A space that is not of the first category is said to be of the **second category.**

At first thought, these category concepts may seem somewhat artificial. (Admittedly, they are not very descriptive.) Once we prove the Baire category theorem, we will have many examples of spaces of the second category. Before doing this, we should see that not all spaces are of the second category by seeing an example of a space that is of the first category. To think of one, observe that a set consisting of a single point is certainly nowhere dense in $\mathbf{R}$ with its usual topology.

3.6. Exercise.
     Give an example of a metric space which is of the first category.

* After René Louis Baire (1874–1932)

If your space is discrete, give one that is not. If your space is not discrete, give one that is.

3.7. Theorem. (The Baire category theorem.) A complete metric space is of the second category.

We can use the Baire category theorem to get an alternate proof of the fact that the product of two Sorgenfrey lines is not a normal space.

3.8. Exercise.

1) Let $X = \mathbf{R}$ with the Sorgenfrey topology and let $X \times X$ have the product topology. Show that $X \times X$ is not normal by showing that the disjoint closed subsets $\{(x, -x) : x$ is rational$\}$ and $\{(x, -x) : x$ is irrational$\}$ cannot be separated by disjoint open sets. Do this by using the fact that the line $\{(x, x) : x \in \mathbf{R}\}$ is of the second category.

2) As another rather interesting example of a completely regular space which is not normal, we have the following. This example is due to V. Niemytzki.

   Let X be the closed upper half plane $\{(x, y) : x, y \in \mathbf{R}$ and $y \geq 0\}$ with topology defined as follows: neighborhoods of points of X above the x-axis have the form $S_r(x, y)$ where $0 < r < y$, and neighborhoods of a point $(x, 0)$ on the x-axis are open disks which are tangent to the x-axis at $(x, 0)$, together with the point $(x, 0)$. The following picture should make this clear. We use X^+ to denote the part of X above the x-axis, $\{(x, y) : x, y \in \mathbf{R}$ and $y > 0\}$.

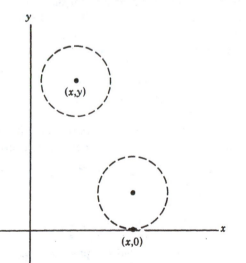

a) Show that X is completely regular by letting F be closed in X and $x_0 \in X - F$. Consider two cases: (1) If $x_0 \in X^+$, use the fact that E^2 is completely regular when given its usual topology to get a continuous (relative to the topology on X!) function from X into I that separates x_0 and F. (2) If x_0 is on the x-axis, then there is a disk D which is tangent to the x-axis at $(x_0, 0)$ and does not meet F. Define a continuous function from X into I which separates x_0 and F. Looking carefully at a picture may help.

b) Show that X is not normal by using a category argument similar to the one in Problem 1 above on an appropriate discrete subspace of X.

c) Does the argument of F. B. Jones (Lemma 4.15 in Chapter 8) apply to show that X is not normal?

4. THE SPACE C*(X)

Let **R** have its usual topology and let $C^*(X)$ be the set of all bounded real-valued continuous functions defined on a metric space X. Define a metric d on $C^*(X)$ by $d(f, g) = \sup_{x \in X} |f(x) - g(x)|$. (This space, with $X = [0, 1]$ was considered in Exercises 7.7 of Chapter 4.)

4.1. Exercise.
 Show that $d(f, g) = \sup_{x \in X} |f(x) - g(x)|$ is a metric on $C^*(X)$. What do r-balls look like in this metric?

It is a fundamental fact of analysis that the space $C^*(X)$ defined above is a complete metric space, although this fact is not always stated in this terminology. To show that $C^*(X)$ is complete, we have to show that every Cauchy sequence of bounded continuous real-valued functions defined on X converges to a bounded continuous real-valued function defined on X. Before doing this, we examine exactly what convergence is in $C^*(X)$.

There are two ways that a sequence of real-valued functions defined on a space can converge: one is called *point-wise* convergence, and the other is called *uniform* convergence. (Compare this with ordinary continuity and uniform continuity.) These are defined as follows.

4.2. Definitions. Let $\{f_n : n \in \mathbf{Z}^+\}$ be a sequence of real-valued functions on a topological space X.

1) The sequence $\{f_n : n \in \mathbf{Z}^+\}$ converges **point-wise** to the function f (written $f_n \to f$, or $\lim_{n \to \infty} f_n = f$) if given $\epsilon > 0$ and $x_0 \in X$, there is an $N \in \mathbf{Z}^+$ such that $|f_n(x_0) - f(x_0)| < \epsilon$ for all $n \geq N$. The function f is called the **point-wise limit** of the f_n's.

2) The sequence $\{f_n : n \in \mathbf{Z}^+\}$ converges **uniformly** to the function f

(written $f_n \rightrightarrows f$ or $f_n \xrightarrow{u} f$) if given $\epsilon > 0$ there is an $N \in \mathbf{Z}^+$ such that $|f_n(x) - f(x)| < \epsilon$ for all $n \geq N$ and for all $x \in X$. The function f is called the **uniform limit** of the f_n's.

The difference between the two methods of convergence, which is similar to the difference between the two forms of continuity, is that in uniform convergence, the rate of convergence is independent of any particular point in the space, while with point-wise convergence, we may have to know which particular point $x_0 \in X$ we are looking at before being able to say how large N must be. In uniform convergence, N depends only on ϵ, while in point-wise convergence, N may depend on both ϵ and x_0. Some examples will make this distinction clear, and we can also see why the distinction is so important.

4.3. Exercises.

1) Let $\{f_n : n \in \mathbf{Z}^+\}$ be a sequence of real-valued functions defined on a topological space X. Show that if $f_n \rightrightarrows f$, then $f_n \to f$.

2) State precisely what it means when a sequence of real-valued functions does not converge uniformly to a function f.

3) The converse to Problem 1 above is false. We consider two examples.

   a) Let $\{f_n : n \in \mathbf{Z}^+\}$ be the sequence of functions from $[0, 1]$ into $\mathbf{R}$ defined by $f_n(x) = x^n$ for each $n \in \mathbf{Z}^+$. Determine a function $f : [0, 1] \to \mathbf{R}$ such that $f_n \to f$. Is f continuous? Does $f_n \rightrightarrows f$?

   b) Let $\{f_n : n \in \mathbf{Z}^+\}$ be the sequence of functions from $[0, 1]$ into $\mathbf{R}$ defined by

   $$f_n(x) = \begin{cases} 2^n x, & \text{if } 0 \leq x \leq \tfrac{1}{2}^n \\ 2 - 2^n x, & \text{if } \tfrac{1}{2}^n \leq x \leq \tfrac{1}{2}^{n-1} \\ 0, & \text{otherwise} \end{cases}$$

   Graph several of the f_n's and determine if there is a function $f : [0, 1] \to \mathbf{R}$ such that $f_n \to f$. Is f continuous? Does $f_n \rightrightarrows f$?

A very important difference between point-wise and uniform convergence is that a sequence of continuous functions can converge point-wise to a non-continuous limit (Exercise 4.3 (3)(a) above), but if the convergence is uniform, then the limit function is also continuous, as you can show in the following theorem. This is really just an application of the triangle inequality and the definition of uniform convergence.

4.4. Theorem. Let $\{f_n : n \in \mathbf{Z}^+\}$ be a sequence of continuous real-valued functions on a topological space X and let f be a continuous real-valued function on X such that $f_n \rightrightarrows f$. Then f is continuous.

In the space $C^*(X)$ with the metic d defined by $d(f, g) = \sup_{x \in X} |f(x) - g(x)|$, it is not hard to show that convergence of a sequence of functions is in fact uniform convergence.

4.5. Theorem. Let $\{f_n : n \in \mathbf{Z}^+\}$ be a sequence of functions in $(C^*(X), d)$. Show that f_n converges to f relative to d (i.e., that given $\epsilon > 0$, there is an $N \in \mathbf{Z}^+$ such that $d(f_n, f) < \epsilon$ for all $n \geq N$) if and only if $f_n \rightrightarrows f$.

Using this result together with the fact that $\mathbf{R}$ is complete, we can prove that $C^*(X)$ is a complete metric space. First we need a result about sequences in r.

4.6. Lemma. Let $\{r_n : n \in \mathbf{Z}^+\}$ be a sequence in $\mathbf{R}$ such that $r_n \to r$. Suppose that there is a number $p \in \mathbf{R}$, a number $B > 0$ and an $N \in \mathbf{Z}^+$ such that $|r_n - p| \leq B$ for all $n \geq N$. Then $|r - p| \leq B$ also.

In other words, if $|r_n - p| \leq B$ for all large n and if $\lim_{n \to \infty} r_n = r$, then $|r - p| \leq B$ also.

The way that this lemma is used in the following theorem is rather subtle. It should be clear where you need it in the proof; then a little thought will tell you how to use it.

4.7. Theorem. The space $C^*(X)$ with metric d^* defined by $d^*(f, g) = \sup_{x \in X} |f(x) - g(x)|$ is a complete metric space.

4.8. Exercise.

(This exercise requires a knowledge of infinite products (in order to know that a point in a product space is a function).)

You might get the impression that since the point-wise limit of a sequence of continuous functions need not be continuous, point-wise convergence is not of much use or interest, and is best avoided. This is not the case at all, because, for example, convergence of a sequence of points in a product space is exactly the same as point-wise convergence of functions.

Let X and Y be topological spaces. For each $x \in X$, let $Y_x = Y$, and let $P = \prod_{x \in X} Y_x$ have the product topology. Then, by definition the set P is the set of all functions (not necessarily continuous) from X into Y. (The set of all continuous functions from X into Y is a subspace of P, and is usually denoted by Y^x.)

Let $\{p_n : n \in \mathbf{Z}^+\}$ be a sequence of points of P. Then each p_n is a function from X to Y. For $p \in P$, show that $p_n \to p$ (relative to the product topology on P) if and only if each $p_n \to p$ (point-wise). In other words, show that $p_n \to p$ in the product topology if and only if for each $x \in X$, $p_n(x) \to p(x)$ in Y.

5. THE COMPLETION OF A METRIC SPACE

As we have seen, not every metric space is complete. We will show in this section that given a non-complete metric space, we can construct a complete metric space that contains it (topologically) as a dense subspace. In other words, if X is a non-complete metric space, we can construct a complete metric space X such that there is a homeomorphism $h: X \to \hat{X}$ and such that $\overline{h(X)} = \hat{X}$. Furthermore, we can do this in such a way that the distance between points in X is the same as the distance between corresponding points in $h(X)$.

A distance preserving homeomorphism is called an **isometry,** and two spaces with an isometry between them are said to be **isometric.** (Recall that not every homeomorphism is an isometry: for example, $(0, 1) \cong \mathbf{R}$.)

Thus, given a non-complete metric space X, we must construct a complete metric space $\hat{X}$ and an isometry $h: X \to \hat{X}$ such that $\overline{h(X)} = X$. (The space $\hat{X}$ so constructed is then called the **completion** of X.) The way that this is done is basically just to "fill in the holes" in X. For example, $\mathbf{Q}$ is not a complete metric space, but $\mathbf{Q}$ is dense in $\mathbf{R}$ (i.e., $\bar{\mathbf{Q}} = \mathbf{R}$), and $\mathbf{R}$ *is* complete. The identity map of $\mathbf{Q}$ into $\mathbf{R}$, given by $h(x) = x$ for $x \in \mathbf{Q}$, is an isometry, and it is true that $\overline{h(\mathbf{Q})} = \mathbf{R}$. In general, of course, things are not so easy, and this example is perhaps even deceptive. The construction that we will give would not produce $\mathbf{R}$ itself as the completion of $\mathbf{Q}$. But it would produce a complete metric space which looks *exactly* like $\mathbf{R}$ topologically. In fact, the only difference between $\mathbf{R}$ and the completion that we will construct is in the way that the points of the space are named, and we know that such a difference is merely superficial.

Incidentally, we have said *the* completion of X as opposed to *a* completion of X, which implies that there is only one completion for a given non-complete metric space. We will show that this is true in the following sense: if X is a non-complete metric space and if X_1 and X_2 are complete metric spaces such that there are isometries (distance preserving homeomorphisms) $h_1: X \to X_1$ and $h_2: X \to X_2$ such that both $\overline{h_1(X)} = X_1$ and $\overline{h_2(X)} = X_2$, then we will show that X_1 and X_2 are isometric. And even more: we will show that there is an isometry h between X_1 and X_2 such that $h(h_1(x)) = h_2(x)$ for all $x \in X$. (Such an isometry is said to *leave the points of X fixed.*)

The construction presented here of the completion of a non-complete metric space is based on the fact that even if X is not complete, $C^*(X)$ is complete (Theorem 4.7), and is adapted from that found in Simmons [14]. To define a homeomorphism of X into $C^*(X)$, we first make some observations. In the rest of this section, let X be a non-complete metric space with metric d, and let d^* be the metric on $C^*(X)$ defined by $d^*(f, g) = \sup_{x \in X} |f(x) - g(x)|$.

We proved in Exercises 1.13 of Chapter 5 that d is a continuous function from $X \times X$ into $\mathbf{R}$. Prove it again here.

5.1. Theorem. The metric $d: X \times X \to \mathbf{R}$ is a continuous function.

Use a straightforward argument involving the triangle inequality to prove the following.

5.2. Theorem. Let $x_0 \in X$. For each $x \in X$, the function f_x defined by $f_x(y) = d(x, y) - d(x_0, y)$ is in $C^*(X)$.

5.3. Theorem. Let $x_0 \in X$. For $x \in X$ define f_x as above. Define $\varphi: X \to C^*(X)$ by $\varphi(x) = f_x$. Then φ is an isometry of X into $C^*(X)$.

When there is an isometry of a metric space X into another metric space Y, we say that X is **imbedded** in Y, or that there is an **imbedding** of X into Y. Thus we have an imbedding of X into $C^*(X)$.

5.4. Theorem. If X is a non-complete metric space, then there is a complete metric space $\hat{X}$ and an isometry $h: X \to \hat{X}$ such that $\overline{h(X)} = \hat{X}$.
[*Hint for proof:* A closed subspace of a complete metric space is complete.]

To show uniqueness of the completion of X, prove the following.

5.5. Theorem. Let Y and Z be complete metric spaces such that X is isometric to a dense subspace of both Y and Z. Then there is an isometry of Y into Z that leaves the points of X fixed (in the sense defined above). [*Hint for proof:* define the isometry of Y into Z by first defining it to be the "identity" on $X"\subset"Y$ onto $X" \subset "Z$. Then extend it to the rest of Y by considering Cauchy sequences in X. (We use quotes because the containments are not literally containments, but are topological containments.)

5.6. Exercise.
   The construction of X given above depends on the particular point $x_0 \in X$ that we choose to begin with. Obviously the construction could be done again using some other point of X, and a different completion of X would result. Show that any two such completions of X are isometric, so that the particular choice of x_0 is immaterial.

Thus, if we have a metric space which is not complete, we know that it can be imbedded in a complete metric space in which it is dense. Furthermore, we know that this can be done so that the space that we started with looks exactly like its image in its completion (except, of course, for the names of the points). As a result, we can regard any metric space as a dense subspace of a complete metric space.

There is another way to construct the completion of a non-complete metric space. This method is perhaps more straightforward than imbedding

X in $C^*(X)$ because it is closer to the "fill in the holes" approach that gives **R** for the completion of **Q**. However, it is more difficult. The basic idea is to adjoin one new point to the space for each Cauchy sequence that does not converge, so that the sequence converges to the new point. An immediate problem with this is that it is possible that two different Cauchy sequences want to converge to the same point, and in such a case, we should add only one new point to the space. This problem is solved and the construction of the completion of a non-complete metric space is outlined in the following exercise. Recall the alternate notation $\langle x_n \rangle$ for a sequence $\{x_n : n \in \mathbf{Z}^+\}$.

5.7. Exercise.
Let (X, d) be a metric space.
1) If $\langle x_n \rangle$ and $\langle y_n \rangle$ are Cauchy sequences in X, we will say that $\langle x_n \rangle$ and $\langle y_n \rangle$ are equivalent, and write $x_n \sim y_n$, if $\lim_{n \to \infty} d(x_n, y_n) = 0$. Let $[\langle x_n \rangle]$ denote the class of all Cauchy sequences in X which are equivalent to $\langle x_n \rangle$.
   a) Prove that the collection of all equivalence classes $[\langle x_n \rangle]$ of Cauchy sequences in X partitions the collection of all Cauchy sequences in X in the sense of Definition 3.4 in Chapter 6.
   b) Prove that if $\langle x_n \rangle$ is a Cauchy sequence in X such that $x_n \to x$, then $\langle y_n \rangle \in [\langle x_n \rangle]$ if and only if $y_n \to x$ also.
2) Let $X^* = \{[\langle x_n \rangle] : \langle x_n \rangle$ is a Cauchy sequence in $X\}$. Make X^* into a metric space as follows. For $\langle x_n \rangle$ and $\langle y_n \rangle \in X^*$, put

$$d^* \left(\langle x_n \rangle, \langle y_n \rangle \right) = \lim_{n \to \infty} d(x_n, y_n).$$

Show that d^* is a metric on X^*. (First of all, you will have to show that d^* is a *function* from $X^* \times X^*$ into **R** (i.e., that d^* is *well-defined*). To do this, you must show that the definition of d^* is independent of the particular representative of each equivalence class that you choose. In other words, show that if $x_n' \sim x_n$ and $y_n' \sim y_n$, then $\lim_{n \to \infty} d(x_n, y_n) = \lim_{n \to \infty} d(x_n', y_n')$.)
3) Show that (X^*, d^*) is complete. (To do this, you are going to have to show that every Cauchy sequence in X^* converges. But the points of X^* are equivalence classes of Cauchy sequences, so you will be working with Cauchy sequences of equivalence classes of Cauchy sequences.)
4) For each $x \in X$, put $\varphi(x) = [\langle x_n \rangle]$ if and only if $x_n \to x$.
   a) Show that φ is a function from X into X^*.
   b) Show that φ is distance preserving, i.e., that

$$d(x, y) = d^*(\varphi(x), \varphi(y)) \text{ for all } x, y \in X.$$

   c) Show that φ is a homeomorphism of X into X^*.

d) Conclude that φ is an isometry of X into X^* and show that $\overline{\varphi(X)} = X^*$. Thus X^* is a complete metric space that contains X (actually, X^* contains an isometric copy of X) as a dense subspace.

e) Prove that X^* is unique up to isometry by showing that if Y and Z are complete metric spaces which both contains X as a dense subspace, then there is an isometry of Y into Z that leaves the points of (the copies of) X fixed. See Theorem 5.5.

6. COMPACTNESS IN METRIC SPACES

We collect some of the results that we have proved in the following theorem.

6.1. Theorem. Let (X, d) be a metric space and let A be a subset of X. The following statements are equivalent.

1) A is compact.
2) A is countably compact.
3) A is sequentially compact.
4) A is pseudocompact.
5) A is complete and totally bounded.

Thus, in any metric space, as in **R**, all five compactness ideas that we have discussed are equivalent.

chapter ten

Compactness Again

We now have several equivalences of compactness for metric spaces and have seen that compact spaces have some nice properties. For example, a continuous real-valued function on a compact set attains both its maximum and its minimum, and a compact Hausdorff space is normal (so that a real-valued continuous function on a closed subset of a compact Hausdorff space can be extended to the whole space). In this chapter, we will see some more properties of compact spaces, and will get a characterization of compactness for general spaces which will enable us to prove the very important theorem that the product of compact spaces is compact. We will also investigate a fascinating and important way of "compactifying" an arbitrary completely regular space in a nice way: we will see how to build a compact space that contains a given completely regular space in such a way that a continuous function from the completely regular space into any compact space can be extended to the compact space that we construct.

1. NETS AND FILTERS

We know that sequences are not adequate when discussing compactness in general, in the sense that a compact space need not be sequentially compact, and a sequentially compact space need not be compact. It is simply not true that every sequence in a compact space must have a convergent subsequence. This is really not the fault of compactness, however; as we know, a sequence with infinitely many distinct terms defines an infinite point-set, and in a compact space, this point-set must have a cluster point. It seems reasonable that some part of the sequence must converge in some sense to such a cluster point, and indeed it does. However, it may not be a subsequence of the sequence which converges to the cluster point. (See

Exercise 4.11 in Chapter 7.) The idea of "something converging in some way" when that "something" is not a sequence will be clearer after we say exactly what it is.

Recall the definition of a totally ordered set (definition 4.2 in Chapter 3). We modify it somewhat to get the following. (In this definition, for "$x < y$" read "x precedes y" or "y follows x.")

1.1. Definition. A set D is a **directed set** if there is an order relation $<$ defined on D which satisfies the following.

1) For $x \in D$, $x < x$ ($<$ is reflexive).
2) For $x, y, z \in D$, if $x < y$ and $y < z$, then $x < z$ ($<$ is transitive).
3) For $x, y \in D$, there is a $z \in D$ such that both $x < z$ and $y < z$.

We say that D is **directed by** $<$, or sometimes, simply that D is **directed.** Instead of $x < y$, we sometimes write $y > x$.

Thus in a directed set D, given any two elements in D, there is a third element in D which follows both of them. Note that this does *not* say that it is always possible to compare the two given elements in any way to each other. In other words, it may happen that given two elements in a directed set D, neither of them precedes or follows the other; however, there must be a third element that follows them both.

1.2. Exercises.

Which of the following sets is directed by the indicated relation?
1) **R** with $\leq$.
2) **R** with $<$.
3) $\mathcal{P}(X)$ for $X \neq \emptyset$ with "$<$" defined to be "is a subset of."
4) $\{(m, n) : m, n \in \mathbf{Z}^+\}$ with $<$ defined by $(m_1, n_1) < (m_2, n_2)$ if and only if both $m_1 \leq m_2$ and $n_1 \leq n_2$.
5) $\mathbf{Z}^+$ with $\leq$.

A sequence in a set X is a function from $\mathbf{Z}^+$ into X. As you saw in Exercise 1.2 (5) above, $\mathbf{Z}^+$ is directed by $\leq$, so a directed set is a generalization of how $\mathbf{Z}^+$ behaves under $\leq$. We will use this fact to generalize sequences.

1.3. Definition. A **net** in a set X is a function from a directed set D into X.

As with sequences, it is important to be aware of the fact that a net is a function, but it is often clearer to write a net in set notation. Thus if $\mathfrak{D}$ is a directed set and $\mathfrak{N} : \mathfrak{D} \to X$ is a net, we will often write $\{x_\alpha : \alpha \in \mathfrak{D}\}$ to denote this net, but we must keep in mind that the net is more than just a set of points.

1. 4. Exercise.

Let $\mathfrak{N} = \{x_\alpha : \alpha \in \mathfrak{D}\}$ be a net. Show that $\mathfrak{D}$ induces a direction on $\mathfrak{N}$ and that $\mathfrak{N}$ itself can be viewed as a directed set.

Thus if $\{x_\alpha : \alpha \in \mathfrak{D}\}$ represents a net in X, then, as with sequences, given x_α and x_β in this net, there is an x_γ in the net such that x_γ is further out in the net than either of x_α or x_β.

As with sequences, nets can converge and they can have accumulation points.

1.5. Definitions. Let $\mathfrak{N} = \{x_\alpha : \alpha \in \mathfrak{D}\}$ be a net in a topological space X, where $\mathfrak{D}$ is directed by $<$.

1) $\mathfrak{N}$ **converges** to x_0 in X if given any neighborhood U of x_0, there is an $\alpha_0 \in \mathfrak{D}$ such that for all $\alpha \in \mathfrak{D}$ with $\alpha > \alpha_0$, $x_\alpha \in U$.

2) A point $x_0 \in X$ is an **accumulation point** of $\mathfrak{N}$ if given any neighborhood U of x_0 and any $\alpha_0 \in \mathfrak{D}$, there is an $\alpha \in \mathfrak{D}$ with $\alpha > \alpha_0$ such that $x_\alpha \in U$.

It should be clear that these two definitions are direct generalizations of the same concepts for sequences. We can even generalize the words "ultimately" and "frequently" to apply to nets.

1.6. Definitions. Let $\mathfrak{N} = \{x_\alpha : \alpha \in \mathfrak{D}\}$ be a net in a set X, where $\mathfrak{D}$ is directed by $<$. Let $S \subset X$.

1) $\mathfrak{N}$ is **ultimately** in S if there is an $\alpha_0 \in \mathfrak{D}$ such that for all $\alpha > \alpha_0$, $x_\alpha \in S$.

2) $\mathfrak{N}$ is **frequently** in S if for all $\alpha_0 \in \mathfrak{D}$ there is an $\alpha \in \mathfrak{D}$ with $\alpha > \alpha_0$ and $x_\alpha \in S$.

Thus we can say that a net in a topological space converges to a point if it is ultimately in every neighborhood of the point, and a point is an accumulation point of a net if the net is frequently in every neighborhood of the point. As with sequences, ultimately implies frequently, but not conversely. In other words, if a net converges to a point, then that point is an accumulation point of the net, but a net can have an accumulation point and still not converge.

Actually, the idea of a net is really not a new one, because you have used it before. Indeed, sequences are not even adequate in beginning calculus, where the only space dealt with is the real line. Recall that the definite integral $\int_a^b f(x)\ dx$ is defined to be a "limit of Riemann sums" in the following sense.

1.7. Exercise.

Let $a, b \in \mathbf{R}$ with $a < b$. A *partition* of $[a, b]$ is a finite set $\{x_0 = a < x_1 < x_2 < \cdots < x_n = b\}$ of points of $[a, b]$. The *norm* of a partition is the length of the largest subinterval $[x_i, x_{i+1}]$ where x_i, x_{i+1} are successive points in the partition. Let $f : [a, b] \to \mathbf{R}$. A *Riemann sum* for f has the form

$$R(f, [a, b], P, x_i^*) = \sum_{i=0}^{n-1} f(x_i^*)\, \Delta x_i$$

where $P = \{x_0, x_1, \cdots x_n\}$ is a partition of $[a, b]$, $x_i^* \in [x_i, x_{i+1}]$ and $\Delta x_i = x_{i+1} - x_i$.

Then $\int_a^b f(x)\, dx$ is defined to be the limiting value of all Riemann sums for f taken over all partitions of $[a, b]$, as the norm of the partitions goes to 0, if this limiting value exists.

This definition is not very precise, because the way that the "limiting value" is to be determined is not specified (and cannot be in terms of convergent sequences alone).

Formulate a precise definition of $\int_a^b f(x)\, dx$ in terms of nets.

We showed (Theorem 2.4 and Exercise 2.6 in Chapter 8) that convergence of sequences is unique in a Hausdorff space, but the fact that convergence of sequences is unique does not necessarily imply that a space is Hausdorff. Using nets, we can avoid this deficiency of sequences and can characterize the Hausdorff property in terms of uniqueness of convergence, as follows.

1.8. Theorem. A topological space X is a Hausdorff space if and only if a net in X can converge to at most one point.

Note that Theorem 1.8 does not say that a space is Hausdorff if and only if every net in the space converges. In fact, it says nothing at all about nets that are not convergent. It *does* say that a space is Hausdorff if and only if every net that already converges must converge to exactly one point.

The key to using nets is the following theorem, which is a generalization of the way that we can construct a sequence converging to a given point in a 1st countable space (and, in particular, in a metric space).

1.9. Theorem. Let $\mathfrak{U} = \{U_\lambda : \lambda \in \Lambda\}$ be a local neighborhood basis at a point x in a topological space X. Then $\mathfrak{U}$ is directed by the relation $\supset$ and there is a net in X that converges to x.

Using Theorem 1.9, we can characterize closure in topological spaces in terms of nets. This generalizes a similar result using sequences in 1st countable spaces.

1.10. Theorem. Let A be a subset of a topological space X. Then a point $x \in X$ is in $\bar{A}$ if and only if there is a net contained in A that converges to x.

In a metric space, a function is continuous if and only if it preserves convergent sequences. This generalizes to the following.

1.11. Theorem. Let X and Y be topological spaces and let $f : X \to Y$. Then f is continuous if and only if whenever $\{x_\alpha : \alpha \in \mathfrak{D}\}$ is a net in X that converges to $x \in X$, then $\{f(x_\alpha) : \alpha \in \mathfrak{D}\}$ converges to $f(x)$ in Y.

Thus a function is continuous if and only if it preserves convergence, but this convergence needs to be more general than just convergent sequences.

We can characterize compactness in terms of nets as follows.

1.12. Theorem. A subset A of a topological space X is compact if and only if every net contained in A has a cluster point (which is a point of A).

There are two more words that are often used in connection with nets.

1.13. Definition. Let D be directed by $<$.
  1) A subset T of D is a **terminal** subset of D (or is **terminal in** D) if there is an $\alpha_T \in D$ such that $T = \{\alpha \in D : \alpha > \alpha_T\}$
  2) A subset C of D is a **cofinal** subset of D (or is **cofinal in** D or is **cofinal with** D) if for every $\alpha_0 \in D$, $C \cap \{\alpha \in D : \alpha > \alpha_0\} \neq \emptyset$.

Thus T is terminal in D if T ultimately equals D, and C is cofinal in D if C meets D frequently.

1.14. Theorem. Let $\mathfrak{N} = \{x_\alpha : \alpha \in \mathfrak{D}\}$ be a net in a space X. Then if $x \in X$,
  1) $x_\alpha \to x$ if and only if given any neighborhood U of x, the set $\mathfrak{N} \cap U$ is terminal in $\mathfrak{N}$.
  2) x is an accumulation point of $\mathfrak{N}$ if given any neighborhood U of x, $\mathfrak{N} \cap U$ is cofinal in $\mathfrak{N}$.

We can use the idea of terminal subsets of a net to obtain another concept of convergence in topological spaces which is really only a variation of the idea of a net. We do this because we want to prove that the product of compact spaces is compact, and it is much easier to do this using this alternate idea of convergence. The following theorem is the key to the definition.

1.15. Theorem. Let $\mathfrak{N} = \{x_\alpha : \alpha \in \mathfrak{D}\}$ be a net in a topological space X and let $\mathfrak{F} = \{T_\alpha : \alpha \in \mathfrak{D}\}$ be the collection of all terminal subsets of $\mathfrak{N}$. Then $\mathfrak{N}$ converges to a point $x \in X$ if and only if given any neighborhood U of x, there is an $\alpha \in \mathfrak{D}$ such that $T_\alpha \subset U$.

We generalize the collection of terminal subsets of a net as follows.

1.16. Definition. A non-empty collection $\mathfrak{F}$ of non-empty subsets of a set X is called a **filter** on X if
  1) Given F_1 and $F_2 \in \mathfrak{F}$, there is an $F_3 \in \mathfrak{F}$ such that $F_3 \subset F_1 \cap F_2$, and

2) If $F \in \mathfrak{F}$ and S is any subset of X such that $F \subset S$, then $S \in \mathfrak{F}$ also.

1.17. Theorem. Let $\mathfrak{F}$ be a filter on X. If $F_1, F_2 \in \mathfrak{F}$ then $F_1 \cap F_2 \in \mathfrak{F}$.

When $A \subset B$, B is called a **superset** of A. Using this idea, we can say that a filter $\mathfrak{F}$ on X is a non-empty collection of non-empty sets that contains all intersections of pairs of elements in $\mathfrak{F}$ and also contains all supersets of members of $\mathfrak{F}$.

1.18. Exercises.

1) Prove that if $\mathfrak{F}$ is a filter on X and $F_1, F_2 \in \mathfrak{F}$ then $F_1 \cap F_2 \neq \emptyset$.
2) If $\mathfrak{F}$ is a filter on X and $F \in \mathfrak{F}$ then $\overline{F} \in \mathfrak{F}$.
3) Prove that if $\mathfrak{F}$ is a filter on X, then $X \in \mathfrak{F}$.
4) Let X be a topological space and let $\mathfrak{F}$ be the collection of all neighborhoods of a particular point $x \in X$. Is $\mathfrak{F}$ a filter on X?
5) Is the collection $\{[n, \infty): n \in \mathbf{Z}^+\}$ a filter on $\mathbf{R}$? Is

$$\{(-n, n): n \in \mathbf{Z}^+\}$$

a filter on $\mathbf{R}$?
6) It $\mathfrak{F}$ is a filter on X and $F_1, F_2, F_3 \in \mathfrak{F}$, is $F_1 \cap F_2 \cap F_3 \in \mathfrak{F}$? Can you generalize this?
7) A filter $\mathfrak{F}$ on a space X is said to be *fixed* if $\cap\{\overline{F}: F \in \mathfrak{F}\} \neq \emptyset$. Otherwise, $\mathfrak{F}$ is said to be *free*.
   a) Give an example of a free filter on $\mathbf{R}$.
   b) Give an example of a fixed filter on $\mathbf{R}$.
   c) Show that if a topological space X is compact, then every filter on X is fixed. [See Problem 6 above.]
   d) Is the coverse to (c) above true?
8) Do the terminal sets of a net satisfy the superset property of filters? Is it necessarily true that if T_α and T_β are terminal sets of a net, then $T_\alpha \cap T_\beta$ is also a terminal set of the net?
9) A collection $\mathfrak{B}$ of subsets of a set X is called a *filterbase* on X if (1) $\emptyset \notin \mathfrak{B}$, and (2) if $B_1, B_2 \in \mathfrak{B}$, then there is a $B_3 \in \mathfrak{B}$ such that $B_3 \subset B_1 \cap B_2$. A filter is then generated by a filterbase by taking the filter to be all members of the filterbase together with all supersets of members of the filterbase. All of the work that we will do with filters could just as well be done with filterbases, with very little modification, and the choice of filters over filterbases is not really a very important one.
   a) Show that the collection of all terminal sets of a net in X is a filterbase on X.
   b) If $\mathfrak{B}$ is a filterbase on X and $B \in \mathfrak{B}$, must $\overline{B} \in \mathfrak{B}$?

Like nets, filters can be used to describe convergence in topological spaces, as follows. (We did this in a special case in Theorem 1.15.)

1.19. Definition. Let $\mathfrak{F}$ be a filter on a topological space X, and let $x \in X$.

1) $\mathfrak{F}$ **converges** to x (written $\mathfrak{F} \to x$ or $\lim \mathfrak{F} = x$) if given any neighborhood U of x, there is an $F \in \mathfrak{F}$ such that $F \subset U$.

   2) $\mathfrak{F}$ **accumulates** at x if given any neighborhood U of x, $F \cap U \neq \emptyset$, for all $F \in \mathfrak{F}$.

1.20. Exercises.
1) Show that if a filter $\mathfrak{F}$ on a space X converges to $x \in X$, then $\mathfrak{F}$ accumulates at x.
2) Give an example of a filter that accumulates at $x \in X$ but does not converge to x.

1.21. Definition. The **adherance** of a filter $\mathfrak{F}$ on a space X is defined to be

$$\text{ad } \mathfrak{F} = \cap \{\bar{F} : F \in \mathfrak{F}\}.$$

1.22. Theorem. Let $\mathfrak{F}$ be a filter on a space X. Then $\mathfrak{F}$ accumulates at $x \in X$ if and only if $x \in \text{ad } \mathfrak{F}$.

Thus a filter $\mathfrak{F}$ on X accumulates at a point $x \in X$ if and only if x belongs to the closure of every member of $\mathfrak{F}$.

We have said that the idea of a filter on X is really only a variation of the idea of a net in X. You can justify this in the following exercises.

1.23. Exercises.
1) Let $\mathfrak{F}$ be a filter on X. Show that $\mathfrak{F}$ is a directed set and use $\mathfrak{F}$ to define a net $\mathfrak{N}$ in X such that if $x \in X$, $\mathfrak{F}$ converges to x if and only if $\mathfrak{N}$ converges to x.
2) Let $\mathfrak{N}$ be a net in X. Use $\mathfrak{N}$ to define a filter $\mathfrak{F}$ on X such that if $x \in X$, then $\mathfrak{N}$ converges to x if and only if $\mathfrak{F}$ converges to x. [*Hint:* See Theorem 1.15.]

The following theorem gives the properties that we have proved for nets in terms of filters. For part of it, we need the following.

1.24. Lemma. Let $\mathfrak{F}$ be a filter on X and let $f : X \to Y$. Then the collection of sets $f(\mathfrak{F}) = \{f(F) : F \in \mathfrak{F}\}$ is a filter on $f(X)$ (but may not be a filter on Y).

1.25. Theorem. Let X and Y be topological spaces, $A \subset X$.
1) X is a Hausdorff space if and only if every filter on X can coverge to at most one point.
2) A point $x \in X$ is in $\bar{A}$ if and only if there is a filter on X that converges to x.

3) A function $f: X \rightarrow Y$ is continuous if and only if whenever $\mathcal{F}$ is a filter on X that converges to $x \in X$, then $f(\mathcal{F})$ converges to $f(x) \in Y$.

4) The set A is compact if and only if for every filter $\mathcal{F}$ on A, there is a point $x \in A$ such that $\mathcal{F}$ accumulates at x. (In other words, A is compact if and only if for every filter $\mathcal{F}$ on A, (ad $\mathcal{F}$) $\cap A \neq \emptyset$.)

As a variation of Theorem 1.25 (3) above, we have the following very nice characterization of continuity in terms of filters.

1.26. Theorem. Let X and Y be topological spaces. A function $f: X \rightarrow Y$ is continuous if and only if for every point $x \in X$, $f(\mathcal{U}(x))$ converges to $f(x)$, where $\mathcal{U}(x)$ is the filter on X consisting of all neighborhoods of x. (The reason for using $\mathcal{U}(x)$ rather than $\mathcal{F}(x)$ will be clear later.)

2. ULTRAFILTERS

A collection of sets with a certain property is said to be **maximal** with respect to the property if it is not a proper subset of a (larger) collection of sets with the same property. For example, the collection $\{S_{1/n}(x): n \in \mathbf{Z}^+\}$ of open neighborhoods of a point x in the metric space $\mathbf{R}$ is not a maximal collection of open neighborhoods of x, because not every open neighborhood of x has this form. On the other hand, this collection *is* maximal with respect to the property of being a collection of open balls of radius $1/n$ that contain x.

The collection of $1/n$-balls of x discussed above is not a filter (Why not?), and at first thought, it might seem that every filter is maximal, because of the superset property. This is false, as you can show in Problem 1 of the exercises below.

2.1. Exercises.

1) Let $\mathcal{F}$ be the collection of all subsets of $\mathbf{R}$ that contain $[0, 1]$. Show that $\mathcal{F}$ is a filter which is not maximal. (In other words, after showing that $\mathcal{F}$ is a filter, exhibit another filter that contains $\mathcal{F}$ as a proper subset.)

2) Let X be a topological space, $x \in X$, and let $\mathcal{F}(x)$ be the collection of all subsets F of X that contain x. Show that $\mathcal{F}(x)$ is a maximal filter on X.

A maximal filter is called an **ultrafilter.** Ultrafilters have some very nice properties.

2.2. Theorem.

1) If $\mathcal{U}$ is an ultrafilter on a space X and $A \subset X$ is such that $A \cap U \neq \emptyset$ for all $U \in \mathcal{U}$, then $A \in \mathcal{U}$.

2) If $\mathcal{U}$ is an ultrafilter on X, then for any $A \subset X$, either $A \in \mathcal{U}$ or $X - A \in \mathcal{U}$.

3) For any point x in a topological space X, is the collection of all neighborhoods of x an ultrafilter on X?

The following property of ultrafilters will be of great importance.

2.3. Theorem. Let X be a topological space and let $\mathfrak{u}$ be an ultrafilter on X. Then $\mathfrak{u}$ accumulates at a point $x \in X$ if and only if $\mathfrak{u}$ converges to x.

Thus ultrafilters behave in the same way as increasing (or decreasing) sequences in **R**, in the sense that if they accumulate at a point, then they must converge to that point.

Combining Theorem 2.3 with (4) of Theorem 1.25, we might claim that we have proved the following.

2.4. Theorem. A topological space is compact if and only if every ultrafilter on X converges.

The theorem is true, but our "proof" would be wrong. The problem comes in the "if part," and is the following. Suppose that we know that every ultrafilter on X converges, and we want to know that X is compact. By Theorem 2.3 we know that since every ultrafilter on X converges, then every ultrafilter on X has an accumulation point. Then, when we try to use Theorem 1.25(4), we find that we need to know that every *filter* (and not just every *ultrafilter*) on X has an accumulation point. This is more than we know.

Probably this faulty reasoning would have been obvious the minute that you tried to use it, and you would not have claimed to have proved Theorem 2.4 in this way in the first place. We mention it because the solution to the problem of finding a correct proof of Theorem 2.4 is very subtle and very deep. In fact, it cannot be done without appealing to a result which is equivalent to the axiom of choice. Rather than become involved in a long digression, we will merely state both the axiom of choice (which we have already used to conclude that the arbitrary product of non-empty sets is non-empty), and the equivalence of it which is most useful for our purposes here. You can find more equivalences of the axiom of choice, as well as the proofs that the various statements are equivalent, in [4] or [9].*

2.5. The Axiom of Choice. Let $\{X_\lambda : \lambda \in \Lambda\}$ be a non-empty collection of non-empty pairwise disjoint sets (i.e., $X_\lambda \cap X_\mu = \emptyset$ for all $\lambda, \mu \in \Lambda$ with $\lambda \neq \mu$). Then there is a function $c : \Lambda \to \bigcup_{\lambda \in \Lambda} X_\lambda$ such that $c(\lambda) \in X_\lambda$ for each $\lambda \in \Lambda$.

* There is a good discussion of these equivalences in [18].

The function c guaranteed by the axiom of choice is often called a **choice function** for $\{X_\lambda : \lambda \in \Lambda\}$: Stated in other words, the axiom of choice says that given any non-empty collection of non-empty sets, then there exists a set consisting of exactly one element from each set in the collection.

To state the equivalence of the axiom of choice that we will use in proving Theorem 2.4, we need some terminology. We have already defined a **partially ordered** set, but, for convenience, we repeat the definition here. The relation $\prec$ is a **partial order** on a set X if (1) $x \prec x$ for all $x \in X$ (so that $\prec$ is reflexive), (2) if for x, $y \in X$, both $x \prec y$ and $y \prec x$, then $x = y$ (so that $\prec$ is anti-symmetric), and (3) if for x, $y, z \in X$, $x \prec y$ and $y \prec z$, then $x \prec z$ (so that $\prec$ is transitive). When $\prec$ is a partial order on X, $(X, \prec)$ is called a **partially ordered set.** (Thus, for example, if S is any non-empty set, $\mathcal{P}(S)$ is partially ordered by $\supset$ and by $\subset$.) A **chain** in a partially ordered set $(X, \prec)$ is a subset C of X such that for x, $y \in C$, either $x \prec y$ or $y \prec x$. (Thus a chain in a partially ordered set $(X, \prec)$ is a totally ordered subset of X.) An **upper bound** for a chain C in a partially ordered set $(X, \prec)$ is an element $x_0 \in X$ such that $c \prec x_0$ for all $c \in C$. Finally, an element $m \in X$ is **maximal** if there is no $x \in X$ with $m \prec x$.

We can now state the following, which is equivalent to the axiom of choice. See, for example, [4] or [9] for a proof.

2.6. Zorn's Lemma.* If every chain in a partially ordered set has an upper bound, then there is a maximal element in the set.

We can use Zorn's lemma to get a result which will enable us to prove Theorem 2.4.

2.7. Theorem. Every filter on a set X is contained in an ultrafilter.
Outline of Proof: Let $\mathcal{F}$ be a filter on X and let $\mathcal{a}$ be the collection of all filters that contain $\mathcal{F}$. Then $\mathcal{a}$ is non-empty because $\mathcal{F} \in \mathcal{a}$. Let $\mathfrak{C} = \{\mathcal{C}_\lambda : \lambda \in \Lambda\}$ be a chain in $\mathcal{a}$ (so the elements $\mathcal{C}_\lambda \in \mathfrak{C}$ are filters on X that contain $\mathcal{F}$.) Show that $\cup \mathfrak{C} = \{C : C \in \mathcal{C}_\lambda$ for some $\lambda \in \Lambda\}$ is a filter on X which is an upper bound for $\mathfrak{C}$. Finish the proof by applying Zorn's lemma to $\mathcal{a}$.

Using Theorem 2.7 you can now prove Theorem 2.4 which we restate here.

2.8. Theorem. A topological space X is compact if and only if every ultrafilter on X converges.

As is often the case with the development of the theory in a text book, we have developed so much powerful machinery that the theorem that we were working toward now falls out as a rather easy corollary. When you

* After Max August Zorn (1906–   ).

prove it, think of all that has been done to get us to this point—filters, ultrafilters, Zorn's lemma, the theory of product spaces and compact spaces.

2.9. Theorem. The product of compact spaces is a compact space.

Theorem 2.9 is called the **Tychonoff product theorem,** after A. Tychonoff, who, incidentally, did not prove it in such generality, but certainly paved the way for such a general statement. It is one of the most important theorems in mathematics.

We can use the Tychonoff theorem to establish some examples that we have deferred until now.

2.10. Exercises.

1) (This exercise requires a knowledge of infinite products.) A compact space need not be sequentially compact. Let I denote the unit interval. For each $i \in I$, let $X_i = I$ (with its usual topology), and let $P = \prod_{i \in I} X_i$ have the product topology.

   a) P is compact, but

   b) P is not sequentially compact. To show it, we will exhibit a sequence in P with no convergent subsequence. To do this, observe that every real number r between 0 and 1 has a (unique) binary decimal expansion $r = 0.d_1 d_2 d_3 \cdots$ with each $d_i = 0$ or 1. Furthermore, every binary decimal $0.d_1 d_2 d_3 \cdots$ with each $d_i = 0$ or 1 represents a (unique) real number between 0 and 1. (The representation is unique if we identify things like 0.0111 and $0.1000 \cdots$. Compare this with the ternary expansion that we used in the discussion of the Cantor set.)

   A point in P can be specified by giving its coordinates. (Actually, $x \in P$ is a function from I to I, and we write $x(t) = x_t$ for $t \in I$.) Let $\{x^n : n \in \mathbf{Z}^+\}$ be the sequence in P such that the t-th coordinate of a term x^n in the sequence is given by $x_t^n = $ n-th coordinate in the binary expansion of t. Suppose that $\{x^{n_k} : k \in \mathbf{Z}^+\}$ is a subsequence of $\{x^n : n \in \mathbf{Z}^+\}$ that converges to $x \in P$. Then $x_k^{n_k} \to x_t$ for each $t \in I$. Consider $t \in I$ with binary expansion given by $0.d_1 d_2 d_3 \cdots$ where

$$d_i = \begin{cases} 0, & \text{if } i = n_k \text{ with } k \text{ odd} \\ 1, & \text{otherwise.} \end{cases}$$

2) (This exercise requires a knowledge of ordinal numbers.) A subspace of a normal space need not be normal. The space $[0, \Omega] \times [0, \Omega]$ is normal (see Problem 8 in Exercises 1.4 of Chapter 7), but $[0, \Omega] \times [0, \Omega)$ is not normal (see Problem 2 in Exercises 4.15 of Chapter 8).

3) (This exercise requires a knowledge of infinite products.) That the product of compact spaces is compact is a good reason for defining the product topology as we did. Show that the product of infinitely many copies of the unit interval with its usual topology is not compact when this product is given the box topology in which the product of open sets is open.

4) Any Euclidean space $E^n = \{(x_1, x_2, \cdots, x_n) : x_i \in \mathbf{R}\}$ (with the product topology where each factor has the usual topology on $\mathbf{R}$) is locally compact.

3. THE STONE-ČECH COMPACTIFICATION

In this section we will investigate a way to compactify a (completely regular) topological space so that a continuous function from the space to a compact (Hausdorff) space can be extended to the compactification. We first review our earlier discussion of compactification, and of the construction of the one-point compactification of a (locally compact) Hausdorff space.

3.1. Definition. Let X be a tolopogical space. A **compactification** of X is a compact space X^* such that there is a homeomorphism $h: X \to X^*$ with $\overline{h(X)} = X^*$.

Thus a compactification of a topological space is a compact space that contains X (topologically) as a dense subspace.

3.2. Theorem. Let X be a locally compact Hausdorff space. Define X^* to be $X \cup \{p\}$, where p is any object which is not an element of X, and topologize X^* as follows: for $x \in X$, neighborhoods of x in X^* are the same as in X, and $U \subset X^*$ is a neighborhood of p if and only if $p \in U$ and $X^* - U$ is a compact subset of X. Then the space X^* is a compact Hausdorff space, and there is a homeomorphism $h: X \to X^*$ such that $\overline{h(X)} = X^*$. The space X^* is called the **one-point compactification** of X.

That the one-point compactification is not totally satisfactory as far as continuous functions are concerned can be seen in the following exercise.

3.3. Exercise.

Let $(0, 1]$ have its usual topology. Show that the one-point compactification of $(0, 1]$ is $[0, 1]$ (up to homeomorphism, of course). Let $f: (0, 1] \to I$ be defined by $f(x) = \sin(1/x)$. Show that f is bounded and continuous, but that it is impossible to extend f over $[0, 1]$.

Thus it may not be possible to extend all continuous functions from a space into a compact space to its one-point compactification. We will build a compactification of a (completely regular) space where such an extension is always possible. Actually, most of the work has already been done. Recall (Theorem 3.14 in Chapter 8) that any completely regular space can be imbedded in a product of unit intervals. We review the procedure in the following theorem.

3.4. Theorem. Let X be a completely regular space and let $C^*(X)$ be the collection of all bounded real-valued continuous functions on X. For each $f \in C^*(X)$, put $I_f = I$, the unit interval with its usual topology, and let $P = \prod_{f \in C^*(X)} I_f$ have the product topology. Define $\varphi: X \to P$ by $\varphi(x) = (f(x))_{f \in C^*(X)}$. Then φ is a homeomorphism of X into P.

Put $\beta X = \overline{\varphi(X)}$. Then βX is a compact space which contains X (topologically) as a dense subspace.

3.5. Definition. The space βX defined in Theorem 3.4 is called the **Stone-Čech* compactification** of X.

As with previous imbeddings, we will regard X as actually being a subset of βX, rather than as just a topological subset. The following theorem is the fundamental characterization of βX.

3.6. Theorem. Let X be a completely regular space, let Y be any compact (Hausdorff) space, and let $F: X \to Y$ be continuous. Then there is a continuous function $F^\beta: \beta X \to Y$ such that $F^\beta | X = F$.

Conversely, if X is contained (topologically or literally) as a dense subspace of a compact space T such that every continuous function from X into a compact (Hausdorff) space Y can be extended to T, then $T \cong \beta X$ by a homeomorphism that leaves the points of (the copies of) X fixed.

Outline of Proof: (The first half of the theorem is known as Stone's theorem; the outline given here is adapted from the proof of Stone's theorem found in Kelley [9].) We have $F: X \to Y$. Define $F^*(g) = g \circ F$ for $g \in C^*(Y)$. Show that $g \circ F$ is a function from X to I so that F^* is a function from $C^*(Y)$ to $C^*(X)$. Define F^{**} by $F^{**}(f) = f \circ F^*$ for each $f \in \beta X$. Show that F^{**} is a continuous function from βX to βY, and show that $\varphi_Y^{-1} \circ F^{**} \circ \varphi_X$ is a continuous extension of F to βX, where φ_X and φ_Y are the imbeddings of X into βX and Y into βY, respectively. The following diagram illustrates the situation.

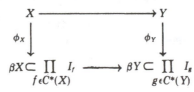

---

* After M. H. Stone and E. Čech, who developed it independently.

For the converse, consider the following diagram.

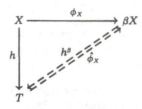

The function h^β is the continuous extension to βX of h which is guaranteed by the first part of the theorem. The function $\hat\phi_x$ is the continuous extension to T of φ_x guaranteed by the hypothesis. Show that $h^\beta \circ \varphi_x = id_T$, and $\varphi_x \circ h^\beta = id_{\beta X}$, and use these facts to show that both h^β and $\hat\phi_x$ are homeomorphisms onto.

When we are working with a completely regular space X, the characterization of βX given in Theorem 3.6 is very convenient if we have a space that we know to be βX, because then we know that every continuous function from X into a compact space Y can be extended to βX. However, if we have a completely regular space X and we want to determine whether or not some compact space actually is βX, then Theorem 3.6 asks us to do a great deal, because we must show that any continuous function on X to *any* compact (Hausdorff) space can be extended to our compact space, before we can conclude that our compact space is βX. (Of course, if our compact space is not βX, then all we have to show to prove it is to exhibit a compact space Y and a continuous $f: X \to Y$ such that f cannot be extended to our compact space.) It is when our compact space *is* βX and we want to prove it that Theorem 3.6 can use some improvement. It would be very helpful if we could reduce the number of compact spaces Y that we need to worry about in such a situation. The following theorem shows that we can reduce the number of Y's considerably.

3.7. Theorem. Let X be a completely regular space that is a dense subspace (topologically or literally) of a compact space T. Let I be the unit interval with its usual topology and suppose that every continuous function from X into I can be extended to T. Then $T \cong \beta X$ by a homeomorphism that leaves the points of X fixed.

[*Hint for proof:* Remember how βX is defined. A diagram may help.]

Thus if X is (topologically) a dense subset of a compact space T and if every bounded real-valued continuous function on X can be extended to T, then $T = \beta X$ (up to homeomorphism).

3.8. Exercises.
(Problems 1 and 2 require a knowledge of ordinal numbers.)

1) Show that $\beta[0, \Omega) = [0, \Omega]$. (See Problem 8 in Exercises 1.4 of Chapter 7.) Thus the one-point compactification can be the same as the Stone-Čech compactification, but as Exercise 3.3 shows, it need not be. Another example is the following.

2) Show that $\beta[0, \omega) \neq [0, \omega]$. In fact, show that $\omega \notin \beta[0, \omega)$.

3) (This is the same as Problem 2 above except that it does not use ordinal numbers.) Let $\mathbf{Z}^+\cup \{\infty\}$ be the one-point compactification of $\mathbf{Z}^+$. (Think of "∞" as being $\lim_{n\to\infty} n$, although it can actually be anything that is not a positive integer. The topology of the one-point compactification puts ∞ at the "end" of the positive integers.) Show that the function $f:\mathbf{Z}^+ \to \mathbf{R}$ defined by $f(n) = 0$ if n is odd, and $f(n) = 1$ if n is even, is continuous and bounded but cannot be extended to ∞. Thus $\beta\mathbf{Z}^+ \neq \mathbf{Z}^+ \cup \{\infty\}$; in other words, the one-point compactification and the Stone-Čech compactification of the positive integers are not the same. Since there is a bounded real-valued continuous function on $\mathbf{Z}^+$ that cannot be extended to ∞, it follows that $\infty \notin \beta\mathbf{Z}^+$ (Where, of course, the position of the point ∞ is determined by the topology on the one-point compactification of $\mathbf{Z}^+$.)

   The problem with ∞ is that it wants to be the limit of every increasing infinite sequence in $\mathbf{Z}^+$ when a neighborhood of ∞ is a set whose complement is compact, as it is in the one-point compactification of $\mathbf{Z}^+$. Certainly, every increasing infinite sequence in $\mathbf{Z}^+$ must have a limit in $\beta\mathbf{Z}^+$ (Why?), but it turns out that they cannot all have the same limit. In other words, for example, $\lim_{n\to\infty} 2n = p_1$ exists in $\beta\mathbf{Z}^+$ and $\lim_{n\to\infty} (2n + 1) = p_2$ exists in $\beta\mathbf{Z}^+$, but it can be shown that $p_1 \neq p_2$. It can also be shown (for example, see [4], p. 244) that in order to be able to extend all bounded real-valued continuous functions to a compactification of $\mathbf{Z}^+$, the compactification must contain 2^c points (where $c = 2^{\aleph_0}$ is the cardinality of $\mathbf{R}$). Thus $|\beta\mathbf{Z}^+| = 2^{2^{|\mathbf{Z}^+|}}$!

4) Let $\mathbf{R}$ have its usual topology. Show that $\beta\mathbf{R}$ is not the same as its one-point compactification. What does the one-point compactification of $\mathbf{R}$ look like?

5) Let X be completely regular. Show that X is connected if and only if βX is connected.

There is another way to construct the Stone-Čech compactification of a given completely regular space which is different from the approach that we took. It builds βX by giving a limit to each ultrafilter on X that does not converge. Chapter 6 of [6] is an excellent account of this approach to βX, and it also gives some interesting properties of the Stone-Čech compactification of several familiar spaces.

chapter eleven

Selected Topics in Point-Set Topology

There are many interesting topological ideas, and we cannot possibly cover all—or even most—of them in a book like this one. In this chapter, we look at a few that we have not considered up to now. Some are extensions of our previous work (like the further discussion of metric spaces, for example), and some involve totally new concepts.

1. MORE ON METRIC SPACES

We have seen that a bounded metric space need not be totally bounded. As a matter of fact, we can show that *any* metric space (X, d) can be given a metric d' which generates the same topology as d, but is such that (X, d') is a bounded metric space, whether (X, d) is or not. Two metrics that generate the same topology on a set are said to be **equivalent metrics.** Thus, given a metric space (X, d), we will produce a metric d' which is equivalent to d and which is bounded. Furthermore, we can make d' as bounded as we like. The key to the definition of d' is that small r-balls in a metric space are enough to generate the metric topology.

1.1. Theorem. Let (X, d) be a metric space. For any positive real number B, there is a metric d' for X which is equivalent to d and is such that $d'(x, y) \leq B$ for all $x, y \in X$.

[*Hint for proof:* For $x, y \in X$, consider min $\{d(x, y), B\}$.]

1.2. Corollary. Given any metric space (X, d), and any positive real number B, there is a metric d' for X which is equivalent to d and is such that $\delta(X) \leq B$ (with the diameter measured using d').

The rest of this section requires a knowledge of infinite products.

Using Theorem 1.1, we can put a metric on the product of countably many metric spaces which generates the product topology. Thus, the countable product of metric spaces is a metric space.

1.3. Theorem. Let $\{(X_n, d_n) : n \in \mathbf{Z}^+\}$ be a countable collection of metric spaces. For each $n \in \mathbf{Z}^+$, let d_n' be a metric for X_n which is equivalent to d_n and is such that $d_n'(x_n, y_n) \leq 1/n^2$ for all $x_n, y_n \in X_n$ Define d on $\prod_{n=1}^{\infty} X_n$ by $d((x_1, x_2, \cdots), (y_1, y_2, \cdots)) = \sum_{n=1}^{\infty} d_n'(x_n, y_n)$. Then d is a metric on $\prod_{n=1}^{\infty} X_n$ and d generates the product topology.

1.4. Exercises.

1) The requirement in Theorem 1.3 that $d(x_n, y_n) \leq 1/n^2$ for each n is used only to ensure that $\sum_{n=1}^{\infty} d_n(x_n, y_n)$ exists. There is another way to approach this problem. Let $\{(X_n, d_n) : n \in \mathbf{Z}^+\}$ be a countable collection of metric spaces. For each $n \in \mathbf{Z}^+$, let d_n' be a metric for X_n such that d_n' is equivalent to d_n, and such that $d_n'(x_n, y_n) \leq 1$ for all $x_n, y_n \in X_n$. Define d' by $d'((x_1, x_2, \cdots), (y_1, y_2, \cdots)) = \sum_{n=1}^{\infty} d_n'(x_n, y_n)/2^n$. Show that d' is a metric on $\prod_{n=1}^{\infty} X_n$ which is equivalent to d (so d' also generates the product topology).

2) For each $n \in \mathbf{Z}^+$, let $X_n = I$, the closed unit interval with its usual topology. Define $I^{\infty} = \prod_{n=1}^{\infty} X_n$. Then I^{∞} is a compact metric space.

The space I^{∞} (sometimes denoted by I^{ω}) is called the **Hilbert cube,**[*] and can also be described as follows. Let X be the set of all sequences $\{x_n : n \in \mathbf{Z}^+\}$ in I such that $0 \leq x_n \leq 1/n$ for each n. Define a metric on X by generalizing the distance formula in the plane: for $x = \{x_n : n \in \mathbf{Z}^+\}$ and $y = \{y_n : n \in \mathbf{Z}^+\}$, put

$$d(x, y) = \left(\sum_{n=1}^{\infty} (x_n - y_n)^2 \right)^{1/2}.$$

That d is a metric on X is easy, except for showing that the triangle inequality holds. We can do this as follows.

We want to show that for sequences $x, y, z \in X$

$$\left(\sum_{n=1}^{\infty} (x_i - y_i)^2 \right)^{1/2} \leq \left(\sum_{n=1}^{\infty} (x_i - z_i)^2 \right)^{1/2} + \left(\sum_{n=1}^{\infty} (z_i - y_i)^2 \right)^{1/2}.$$

If we put $a_i = x_i - y_i$ and $b_i = z_i - y_i$, we then have to show that

$$\left(\sum_{n=1}^{\infty} (a_i + b_i)^2 \right)^{1/2} \leq \left(\sum_{n=1}^{\infty} a_i^2 \right)^{1/2} + \left(\sum_{n=1}^{\infty} b_i^2 \right)^{1/2}.$$

This will be easier to deal with if we use the idea of the *norm* of a sequence in X and write, for $x \in X$, $\| x \| = \left(\sum_{n=1}^{\infty} x_i^2 \right)^{1/2}$. Thus we have to show that $\| a + b \| \leq \| a \| + \| b \|$. We proceed in two steps.

_____
[*] After David Hilbert (1862–1943).

a) (Cauchy's inequality.) For $a, b \in X$ and $N \in \mathbf{Z}^+$,

$$\sum_{i=1}^{N} a_i b_i \leq \| a \| \, \| b \|.$$

To prove this, observe that $\| x \| = 0$ if and only if each $x_i = 0$ (in which case we say $x = 0$). Treat the case $a = 0$ or $b = 0$ as a special case. If neither a nor b is 0, show that for each i,

$$\frac{a_i b_i}{\| a \| \, \| b \|} \leq \tfrac{1}{2} \left(\frac{a_i^{\,2}}{\| a \|^2} + \frac{b_i^{\,2}}{\| b \|^2} \right).$$

[*Hint:* For any real numbers r and s, $(r - s)^2 \geq 0$, so $\sqrt{rs} \leq \tfrac{1}{2}(r + s)$. Add to get Cauchy's inequality.]

b) Show that for each $N \in \mathbf{Z}^+$

$$\sum_{i=1}^{N} (a_i + b_i)^2 \leq \| a + b \| \, (\| a \| + \| b \|).$$

Show that $(X, d) \cong I^\infty$.

3) We can generalize the Hilbert cube to get a complete metric space (H, d) (called **Hilbert space***) in which the closed unit balls, $\bar{S}_1(x) = \{ y \in H : d(x, y) \leq 1 \}$, are not compact. (This should be rather surprising. In R and E^2, the closed unit balls are certainly compact; in $\mathbf{Q}$ they are not, but then $\mathbf{Q}$ is not complete.)

Let H be the set of all sequences $\{x_n : n \in \mathbf{Z}^+\} \subset \mathbf{R}$ such that $\sum_{n=1}^{\infty} x_n^2$ converges (so H consists of all "square summable sequences"). Define a metric d on H by $d((x_1, x_2, \cdots), (y_1, y_2, \cdots)) = \left(\sum_{n=1}^{\infty} (x_n - y_n)^2 \right)^{1/2}$.

a) Show that (H, d) is a complete metric space. (See Problem 2 above. If $\{x^m : m \in \mathbf{Z}^+\}$ is a Cauchy sequence in H, then for each n, $\{x_n^m : m \in \mathbf{Z}^+\}$ is a Cauchy sequence in $\mathbf{R}$ (Why?).)

b) Let $\bar{S}_1(x) = \{ y \in H : d(x, y) \leq 1 \}$. Show that $\bar{S}_1(x)$ is not sequentially compact (and therefore, is not compact, since (H, d) is a metric space). Do this by considering a sequence in $\bar{S}_1(x)$ whose n-th term is a sequence whose n-th term is on the boundary of $\bar{S}_1(x)$, while all of its other terms coincide with x in the center.

c) Prove that $(H, d) \cong \mathbf{R}^\infty$ (where $\mathbf{R}^\infty$ is the product of countably many copies of $\mathbf{R}$, with the product topology. $\mathbf{R}^\infty$ is a metric space by Theorem 1.3 or Problem 1 in Exercises 1.4).

4) The product of uncountably many non-trivial metric spaces is not a metric space. [*Hint:* A metric space must be 1st countable.]

---

* A space that satisfies certain properties is a Hilbert space. This is a special case called simply Hilbert space.

2. METRIZABILITY

As we have seen, not every topological space is a metric space. A topological space is said to be **metrizable** if a metric can be defined on it which generates the topology that is already on the space. Thus, a metric space is certainly metrizable. Not every space is. Much work has been done and several important theorems concerning metrizability have been established. We will look at only one of these, a classic result due to P. Urysohn. First, do the following exercises.

2.1. Exercises.
1) Let X be any topological space. Show that it is always possible to define a metric on X, but this metric may not generate the topology that X already has.
2) Show that every metric space is Hausdorff. Give an example of a topological space that is not metrizable.
3) (This exercise requires a knowledge of ordinal numbers.)
    a) Show that every metric space is 1st countable. Show that $[0, \Omega]$ with its usual topology is not metrizable.
    b) The space $[0, \Omega)$ is 1st countable, and yet it too is not metrizable. However, the proof of this fact is beyond the scope of this book.

The following theorem is known as **Urysohn's metrization theorem.** To prove it, show that the conditions in the hypothesis allow you to imbed the space in I^∞. See Section 1 of this chapter, and Section 3 of Chapter 8.

2.2. Theorem. A regular 2nd countable (Hausdorff) topological space is metrizable.

That the condition in Urysohn's metrization theorem that the space be 2nd countable is not necessary will be obvious if you exhibit a metric space which is not 2nd countable.

2.3. Exercise.
Give an example of a metric space which is not 2nd countable.

3. QUOTIENT SPACES

In this section, we will look briefly at a way to put a topology on a set Y when we have a space X and a function f from X onto Y. The topology that we choose, called the **quotient topology** (also called the **identification topology**) can then be used to identify certain points in a space to get a new space. For example, if we identify the points 0 and 1 in $[0, 1]$, the resulting space is then a circle in the plane.

3.1. Exercise.
Let X and Y be topological spaces. Show that the product topology

on $X \times Y$ is the smallest topology that makes the projection functions continuous. In other words, if $\mathfrak{I}$ is the product topology on $X \times Y$ and if $\mathfrak{I}'$ is a topology on the set $X \times Y$ such that both $\pi_1: X \times Y \to X$ and $\pi_2: X \times Y \to Y$ are continuous relative to $\mathfrak{I}'$, then $\mathfrak{I} \subset \mathfrak{I}'$.

The product topology is the smallest topology on $X \times Y$ that makes the projections continuous. Suppose that X is a topological space, Y is a set and f is a function from X onto Y. We want to give Y a topology so that f is continuous. If we copy the product topology and consider the smallest topology on Y such that f is continuous, we get a very uninteresting space.

3.2. Exercise.

Show that if X is a topological space and f is a function from X onto Y, then the smallest topology on Y that makes f continuous is the indiscrete topology.

Thus the smallest topology that we can give to Y so that f is continuous is trivial, and does not depend on X or f. However, the largest such topology may be of more interest, because in general it will depend both on the topology on X and the function f.

3.3 Exercise.

Let X be a topological space, let Y be a set and let f be a function from X onto Y. Show that the discrete topology on Y is the largest topology that can be given to Y, but that f need not be continuous when Y has the discrete topology.

To get the largest topology on Y that makes a function f from X onto Y continuous, we simply force f to be continuous, as follows.

3.4. Definition. Let X be a topological space, let Y be a set and let f be a function from X onto Y. The **quotient topology** (or the **identification topology**) on Y determined by f and X is the largest topology on Y that makes f continuous, and consists precisely of those sets $U \subset Y$ such that $f^{-1}(U)$ is open in X.

3.5. Exercise.

Show that if X is a topological space, Y is a set and f is a function from X onto Y, then the quotient topology on Y determined by f and X actually is a topology on Y.

If Y already has a topology and f is a continuous function from X onto Y, it is actually rather unusual for the topology on Y to be the quotient topology.

3.6. Exercise.

Give an example of topological spaces X and Y and a continuous function f from X onto Y such that the topology on Y is not the quotient topology determined by f and X.

An important special case when the topology on Y is the quotient topology determined by f and X is when f is either an open or closed function. We repeat the definitions of open and closed functions here. See Exercises 3.5 and 5.5 in Chapter 4.

3.7. Definition. Let X and Y be topological spaces and let $f\colon X \to Y$ be a (not necessarily continuous) function of X into Y. Then

1) f is **open** (an **open function**, or, if f is continuous, an **open map**) if $f(U)$ is open in Y whenever U is open in X.

2) f is **closed** (a **closed function**, or, if f is continuous, a **closed map**) if $f(F)$ is closed in Y whenever F is closed in X.

3.8. Theorem. Let X and Y be topological spaces and let f be a continuous function from X onto Y. If f is either open or closed, then the topology on Y is the quotient topology determined by f and X.

3.9. Exercises.

1) Let X and Y be topological spaces and let $X \times Y$ have the product topology. What is the quotient topology on X determined by $X \times Y$ and π_1?

2) Theorem 3.8 says that if a continuous function from a space X onto a space Y is an open function or a closed function, then Y has the quotient topology determined by f and X. We should note that the converse to this theorem is false. Let $I = [0, 1]$ have its usual topology and define $f\colon I \to \{0, 1\}$ by

$$f(x) = \begin{cases} 0, & \text{if } 0 \le x < \tfrac{1}{2} \\ 1, & \text{if } \tfrac{1}{2} \le x \le 1. \end{cases}$$

Show that the quotient topology on $\{0, 1\}$ determined by f and I is the Sierpinski topology (Example 2.3 in Chapter 4). Show that when $\{0, 1\}$ has the Sierpinski topology (i.e., the quotient topology determined by f and I), the function f is not open.

3) Let X be a topological space and let $A \subset X$ be a retract of A. Let $r\colon X \to A$ be a retraction of X onto A. Then A has the quotient topology determined by r and X.

We can use the quotient topology to get new spaces from old ones. The procedure is to define a set Y to be some modification of a space X, define a function from X onto Y, and give Y the quotient topology determined by this function and X. Since the function is then continuous (because Y has the quotient topology) we may have some information about Y immediately. For example, if X is compact or connected, so is Y. Before we can do this, we need to establish a method of modifying a space X by identifying some of its points.

3.10. Definition. Let E be a relation on a set X. If $x, y \in X$ are related by E, we write xEy. (For example, if E is $<$ on $\mathbf{Z}^+$, we would write xEy to mean $x < y$.) The relation E on X is an **equivalence relation** on X if it satisfies the following.

1) For all $x \in X$, xEx (E is *reflexive*).
2) For all $x, y \in X$, if xEy then yEx (E is *symmetric*).
3) For all $x, y, z \in X$, if xEy and yEz, then xEz (E is *transitive*).

3.11. Exercises.

Which of the following relations is an equivalence relation on the indicated set?

1) $=$ on $\mathbf{R}$.
2) $<$ on $\mathbf{R}$.
3) $\leq$ on $\mathbf{R}$.
4) $\sim$ on $\mathbf{R}$ defined by $x \sim y$ if and only if $x - y \in \mathbf{Q}$.
5) $\cong$ on topological spaces.

Recall (Definition 3.4 in Chapter 6) that a collection of subsets *partitions* a set X if the union of the collection is all of X and the members of the collection are pairwise disjoint.

3.12. Theorem. An equivalence relation E on a set X partitions X into pairwise disjoint subsets called **equivalence classes** (under E), which are defined as follows. The equivalence class (under E) containing $x \in X$ is $[x] = \{y \in X : xEy\}$.

Conversely, given a collection of subsets of a set X that partitions X into pairwise disjoint subsets, we can define an equivalence relation on X, which we denote by $\sim$, by $x \sim y$ if and only if x and y belong to the same subset of the partition.

We can use the idea of an equivalence relation to identify certain points in a topological space (in other words, to make certain points equivalent—the same). For example, define $\sim$ on $[0, 1]$ by $0 \sim 1$ and $x \sim x$ for all $x \in [0, 1]$. By identifying the points 0 and 1 (by making them equivalent), we "paste" the interval together at its end points and obtain a circle.

To relate this to quotient spaces, let X be a topological space and let $\sim$ be an equivalence relation on X. Let $[x]$ denote the equivalence class of $x \in X$ under $\sim$: $[x] = \{y \in X : x \sim y\}$. Let $X/\sim$ denote the set of all distinct equivalence classes under $\sim$ and define a function p from X onto $X/\sim$ by $p(x) = [x]$. (The function p is called the *projection* of X onto the *quotient* $X/\sim$.) Finally, let $X/\sim$ have the quotient topology determined by p and X.

The resulting topological space is called the *quotient space* of X by $\sim$. To summarize this for later reference,

3.13. Definition. Let X be a topological space, $\sim$ an equivalence relation on X and let p be the projection of X onto $X/\sim = \{[x]:x \in X\}$, defined by $p(x) = [x]$. The set $X/\sim$ with the quotient topology determined by p and X is called the **quotient space** of X by $\sim$.

3.14. Exercises.
1) Show that a circle can be viewed as a quotient space of a closed interval as follows.
   a) Define $\sim$ on $[0, 1]$ by $0 \sim 1$ and $x \sim x$ for $x \in [0, 1]$. Show that $[0, 1]/\sim$ is a circle.
   b) Let $C = \{(x, y) \in E^2:x^2 + y^2 = 1\}$ be the unit circle in the plane and let I be the closed unit interval. Let both C and I have their usual topologies. Define $f:I \to C$ by $f(t) = (\cos t, \sin t)$. Show that C has the quotient topology determined by f and I.
2) Let **R** have its usual topology and define $\sim$ on **R** by $x \sim y$ if and only if $x - y \in$ **Z**. What is **R**$/\sim$?
3) Let **R** have its usual topology and define $\sim$ on **R** by $x \sim y$ if and only if $x - y \in$ **Q**. What is **R**$/\sim$?
4) Let **R** have its usual topology and define $\sim$ on **R** by $x \sim y$ if and only if $x = y$. What is **R**$/\sim$?

As Exercise 3.14 (3) above shows, a quotient space obtained from a Hausdorff space need not be Hausdorff. Since we want all of our spaces to be Hausdorff, this presents us with a problem. We can remedy it as follows. Recall (Section 2 of Chapter 8) that a space X is Hausdorff if and only if the diagonal $\{(x, x) \; x \in X\}$ is closed in $X \times X$, and that if $f:X \to Y$ is continuous and Y is Hausdorff, then the set $\{(x_1, x_2):f(x_1) = f(x_2)\}$ is closed in $X \times X$.

3.15. Theorem. Let X be a (Hausdorff) topological space and let f be an open, continuous function from X onto Y. Then Y is a Hausdorff space if and only if the set $\{(x_1, x_2):f(x_1) = f(x_2)\}$ is closed in $X \times X$.

3.16. Exercises.
We can represent several important topological 2-manifolds as quotient spaces of a rectangle as follows. Let S be the rectangle in the plane given by $S = \{(x, y) \in E^2:0 \leq x \leq 2\pi, 0 \leq y \leq 1\}$.
1) *The cylinder.*
   a) If we define $\sim$ on S by $(x, 0) \sim (x, 1)$ for all $x \in [0, 2\pi]$ and $(x, y) \sim (x, y)$ for all points in S, show that $S/\sim$ is a cylinder in E^3. This can be visualized as follows.

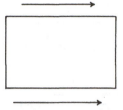

The arrows indicate that we are identifying the two horizontal edges with each other in the same direction.

b) Define $f: S \to E^3$ by $f(x, y) = (\cos x, \sin x, y)$. Show that $f(S)$ is a cylinder when given the quotient topology determined by f and S.

2) *The torus.* Define $\sim$ on S by identifying both pairs of opposite edges of S in the same direction, as in the following picture.

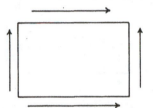

What is $S/\!\!\sim$? Show that it is homeomorphic to $S^1 \times S^1$, where S^1 is a circle in the plane with its usual topology.

3) *The Möbius strip.* Define $\sim$ on S by identifying one pair of opposite edges in the opposite direction, as follows.

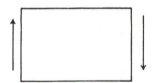

The space $S/\!\!\sim$ is called a *Möbius strip.** What does it look like? Recall that the boundary of a topological 2-manifold is the set of its points that do not have a neighborhood homeomorphic to a neighborhood in E^2, but do have a neighborhood homeomorphic to a neighborhood of a point on the x-axis in the closed upper half plane. What does the boundary of a Möbius strip look like?

4) *The Klein bottle.* Define $\sim$ on S by identifying one pair of opposite edges in the same direction and the other in the opposite direction, as follows.

* After August Ferdinand Möbius (1790–1868).

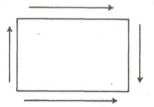

The space $S/\sim$ is called a *Klein bottle*.* What does it look like? Demonstrate that a Klein bottle can be obtained by "pasting" two Mobius strips together along their boundaries.

5) *The projective plane.* Define $\sim$ on S by identifying each pair of opposite edges of S in opposite directions, as follows.

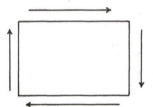

The space $S/\sim$ is called the *projective plane*.

a) Show that the projective plane is homeomorphic to the quotient space of a (hollow) sphere (denoted S^2) in E^3 obtained by identifying diametrically opposite points, as follows.

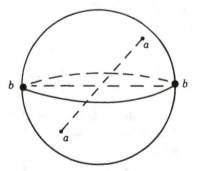

[*Hint:* Blow S up into a hemisphere.]

b) Show that the projective plane is homeomorphic to the space P^2 of all lines L in E^3 through the origin, with topology generated by the basis consisting of elements of the form $B_r(L) \subset P^2$ such that $B_r(L) \cap S^2 \subset S_r(x, y, z) \cap S^2$ where S^2 is the unit sphere

* After Felix Klein (1849–1925).

$\{(x, y, z) \in E^3 : x^2 + y^2 + z^2 = 1\}$ in E^3, and (x, y, z) is on L and on S^2.

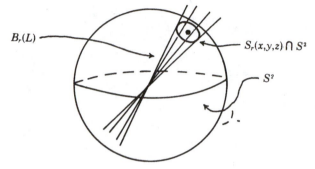

chapter twelve

Homotopy and the Fundamental Group

In this chapter, we investigate very briefly, and, at times, intuitively, a way of classifying topological spaces into classes so that all of the spaces in a given class share some important properties. We already have such a classification theorem if we group homeomorphic spaces into the same class. The classification system that we study here is more general.

The material in this chapter comprises a small part of what is called *algebraic topology*. As the name implies, algebraic topology uses algebraic techniques to discuss topological ideas. We will not go very deeply into the subject, and you will only need to know some basic group theory to work through the material presented here.

Basically, what we will do is to discuss a way of associating a group (called the **fundamental group** or the **first homotopy group**) with a topological space in such a way that the group can tell us something about the space. Let I be the unit interval $[0, 1]$ with its usual topology.

1. HOMOTOPY

Consider the following picture.

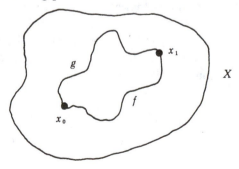

In the picture, X is a topological space, x_0, $x_1 \in X$, and f and g are continuous functions from I into X, with $f(0) = g(0) = x_0$ and $f(1) = g(1) = x_1$. It should be clear that we can deform the graph of f without tearing it or breaking it so that it can be made to coincide exactly with the graph of g. Furthermore, we can do this in such a way that the end points (x_0 and x_1) remain fixed throughout the deformation. Such a change of f into g takes time, and this is the key to making the idea precise. Think of time as running from 0 to 1. We can perform the deformation of f into g in this unit of time by having f at time 0, continuously changing f as time goes from 0 to 1, and finally having g at time 1. This is illustrated below.

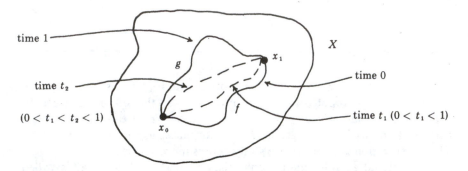

Performing this deformation of f into g suggests the following definition. Recall that a continuous function from I into a topological space X is called a *path* in X with *initial point* $f(0)$ and *terminal point* $f(1)$. Notice that the definition below involves the functions f and g themselves, and not their graphs: we define a deformation of one *function* into another, rather than of the graph of one function into the graph of another. (But, of course, when one function is deformed into another, so is its graph.) This fits with the idea of a path, because a path is a function, and not its graph. (This is so that the idea of the *direction* of a path makes sense: we want to be able to say that a path goes from its initial point to its terminal point.)

1.1. Definition. Let X be a topological space and let f and g be paths in X with common initial points and common terminal points (i.e., $f(0) = g(0)$ and $f(1) = g(1)$). We say that f and g are **path homotopic**, and write $f \underset{p}{\simeq} g$, if there is a continuous function $H : I \times I \to X$ (called a **path homotopy** between f and g) such that $H(s, 0) = f(s)$ for all $s \in I$, $H(s, 1) = g(s)$ for all $s \in I$, $H(0, t) = x_0$ for all $t \in I$, and $H(1, t) = x_1$ for all $t \in I$. This definition is illustrated below.

Notice in the illustration that for "time" t_0, $0 < t_0 < 1$ (on the vertical copy of I), $H(s, t_0)$ is a path in X from x_0 to x_1 which is "between" f and g.

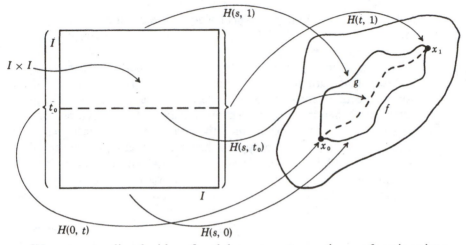

We can generalize the idea of path homotopy to continuous functions into a topological space that are not necessarily paths (or to paths that do not necessarily have common end points). The idea is the same: we deform one function into another in a unit of time, but we drop the requirements that the domain is I and that the end points remain fixed throughout the deformation.

1.2. Definition. Let X and Y be topological spaces and let f and g be continuous functions from X into Y. We say that f and g are **homotopic,** and write $f \simeq g$, if there is a continuous function $H: X \times I \to Y$ (called a **homotopy** between f and g) such that $H(x, 0) = f(x)$ for all $x \in X$ and $H(x, 1) = g(x)$ for all $x \in X$.

Even though $X \neq I$ necessarily in the definition of a homotopy, we still think of $X \times I$ as a square and illustrate the definition of homotopy as we did with path homotopy:

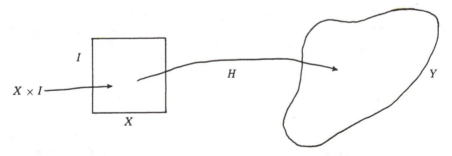

Such a picture is often helpful in deciding how to define a particular homotopy.

1.3. Exercise.

A constant path in a topological space Y is a constant function from I into Y. In other words, $f:I \rightarrow Y$ is a constant path if there is a $y_0 \in Y$ such that $f(s) = y_0$ for all $s \in I$.

Let Y be a topological space. Prove that every path in Y is homotopic to a constant path. (This does *not* say that every path in Y is *path* homotopic to a constant path; indeed, such a statement is almost never true. When is it true?) To show that every path is homotopic to a constant path, observe that since the end points of a path need not remain fixed under a homotopy (as they must under a path homotopy), then if f is a path in Y with $f(0) = y_0$ and $f(1) = y_1$, we can merely push y_1 along the graph of f until it coincides with y_0. This involves shrinking the graph of f down to a point. To get a homotopy to do this, we appeal to a picture for an idea.

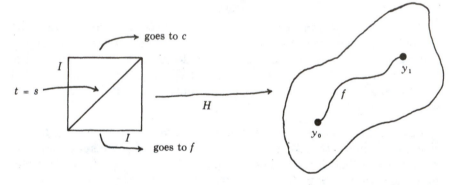

We want the bottom of the rectangle $I \times I$ to be f, and the top to be the constant path $c:I \rightarrow Y$ given by $c(s) = y_0$ for all $s \in I$. And we want the change from f to c to be continuous as we go up the square (i.e., as time goes from 0 to 1). To define a homotopy between f and c, draw the line $t = s$ on the square (the s-axis is the horizontal axis, and the t-axis (the time axis) is the vertical axis). Below this line, we want H to be f, and above the line, we want H to be c. It should be clear that less and less of each of the horizontal I's in the square is f as we go up (i.e., as t goes from 0 to 1). The effect is to shrink the graph of f to a point.

Define H by

$$H(s, t) = \begin{cases} f(s - t), & \text{if } s \geq t \\ y_0 & \text{if } s < t. \end{cases}$$

Show that $H(s, 0) = f(s)$, $H(s, 1) = y_0$ and that on the line $t = s$ where the "change" occurs, H is well defined (i.e., if $s = t$, $f(s - t) = y_0$).

The only problem is whether or not H is continuous. We can prove a general result to establish this.

1.4. Lemma. Let A and B be subsets of a topological space X such that $X = A \cup B$, let Y be a topological space and let $f: A \to Y$, $g: B \to Y$ be continuous functions such that $f(x) = g(x)$ for all $x \in A \cap B$. Define $F: A \cup B \to Y$ by

$$F(x) = \begin{cases} f(x), & \text{if } x \in A \\ g(x), & \text{if } x \in B. \end{cases}$$

If A and B are either both open or both closed in X, then F is a continuous function from $X = A \cup B$ into Y.

1.5. Exercise.

Finish showing that every path in a topological space is homotopic to a constant path.

Recall that an *equivalence relation* on a set is a relation between the elements of the set that is reflexive, symmetric and transitive, and that an equivalence relation on a set partitions the set into pairwise disjoint subsets called *equivalence classes*. Let Y^X denote the set of all continuous functions of a topological space X into a topological space Y.

1.6. Theorem. Homotopy is an equivalence relation on Y^X.

[*Hint for proof:* Reflexivity is easy. For symmetry, if $f \simeq g$, turn the homotopy around to get $g \simeq f$ (see Section 4 in Chapter 6 and draw a picture). For transitivity, if $f \simeq g$ and $g \simeq h$, perform each homotopy in half the time to get $f \simeq h$ (see Lemma 4.5 in Chapter 6, draw a picture and use Lemma 1.4).]

We can partition the collection of topological spaces into pairwise disjoint classes, where each class consists of all of those spaces that have the same *homotopy type*, as defined below.

1.7. Definition. Let X and Y be topological spaces. Then X and Y are said to be **of the same homotopy type** or to **have the same homotopy type** if there are continuous functions $f: X \to Y$ and $g: Y \to X$ such that $g \circ f \simeq id_X$ and $f \circ g \simeq id_Y$.

Before proving that having the same homotopy type is an equivalence relation on the collection of topological spaces (so that it partitions them into pairwise disjoint equivalence classes), consider the following exercises.

1.8. Exercises.

1) Homeomorphic spaces are of the same homotopy type. [*Hint:* "=" implies "$\simeq$".]

2) There exist topological spaces which have the same homotopy type but which are *not* homeomorphic. To show it, let $(x_0, y_0) \in E^2$, and let E^2 have its usual topology. Show that the spaces $\{(x_0, y_0)\}$ and

E^2 have the same homotopy type, but are not homeomorphic. [*Hint:* Show that the identity function on E^2 is homotopic to the constant function that takes E^2 to (x_0, y_0). To do this, consider a line in the plane joining an arbitrary point (x, y) to (x_0, y_0).]

We will have to wait until later to see an example of two spaces that do not have the same homotopy type.

That the relation of having the same homotopy type is an equivalence relation on the collection of all topological spaces is easy except for the transitivity. We can establish transitivity as follows. If X and Y have the same homotopy type, then there are continuous functions $f_1: X \to Y$ and $g_1: Y \to X$ such that $g_1 \circ f_1 \simeq id_X$ and $f_1 \circ g_1 \simeq id_Y$. Similarly, if Y and Z have the same homotopy type, then there are continuous functions $f_2: Y \to Z$ and $g_2: Z \to Y$ such that $g_2 \circ f_2 \simeq id_Y$ and $f_2 \circ g_2 \simeq id_Z$. We want $(g_1 \circ g_2) \circ (f_2 \circ f_1) \simeq id_X$ and $(f_2 \circ f_1) \circ (g_1 \circ g_2) \simeq id_Z$, in order to see that X and Z have the same homotopy type. It would be nice if we could do the following.

$$(g_1 \circ g_2) \circ (f_2 \circ f_1) = g_1 \circ (g_2 \circ f_2) \circ f_1 \simeq g_1 \circ id_Y \circ f_1 = g_1 \circ f_1 \simeq id_X.$$

The problem comes in the middle. Is it true that when $g_2 \circ f_2 \simeq id_Y$ then $g_1 \circ (g_2 \circ f_2) \circ f_1 \simeq g_1 \circ id_Y \circ f_1$? The answer is yes, as the following theorem shows.

1.9. Theorem. If F_1 and F_2 are functions from X into Y with $F_1 \simeq F_2$ and G_1 and G_2 are functions from Y into Z with $G_1 \simeq G_2$, then $G_1 \circ F_1 \simeq G_2 \circ F_2$. [*Hint for proof:* Show that $G_1 \circ F_1 \simeq G_1 \circ F_2$ and then that $G_1 \circ F_2 \simeq G_2 \circ F_2$.]

1.10. Theorem. Same homotopy type is an equivalence relation on the collection of topological spaces.

We saw in Exercise 1.8 (2) that the identity function on E^2 is homotopic to a constant function from E^2 into E^2. Spaces with this property are important because, as far as homotopy is concerned, they are the same as a single point space.

1.11. Definition. A topological space in which the identity function is homotopic to a constant function is said to be a **contractible** space.

1.12. Theorem. A topological space is contractible if and only if it has the same homotopy type as a single point space.

Because of Theorem 1.12, we say that a contractible space is *homotopically trivial*.

1.13. Exercises.
1) The real line and the Euclidean plane, both with their usual topologies, are contractible.

2) The closed unit interval in **R** and the closed unit disk in E^2, both with their usual topologies, are contractible.
3) The open unit interval in **R** and the open unit disk in E^2, both with their usual topologies, are contractible. [*Hint:* See Exercise 1.8 (1).]

2. THE FUNDAMENTAL GROUP

For completeness, we recall the definition of a group here.

2.1. Definition. A group G is a set together with an operation (which we denote here by $*$) satisfying the following.
1) For all g_1, $g_2 \in G$, $g_1 * g_2 \in G$ ($*$ is **closed**).
2) For all g_1, g_2, $g_3 \in G$, $g_1 * (g_2 * g_3) = (g_1 * g_2) * g_3$ ($*$ is **associative**).
3) There is an element $e \in G$ such that for all $g \in G$, $g * e = e * g = g$ (G contains an **identity element** for $*$).
4) If $g \in G$, there is an element $g^{-1} \in G$ such that $g * g^{-1} = g^{-1} * g = e$ (each element in G has an **inverse** in G relative to $*$).

2.2. Exercises.
Which of the following sets and operations is a group?
1) $\mathbf{Z}^+$ with $+$.
2) $\mathbf{Z}$ with $+$.
3) $\mathbf{R}$ with $+$.
4) $\mathbf{R}$ with multiplication.

Recall that a *loop* in a topological space is a path in the space whose initial point and terminal point are the same. If the initial point and terminal point of a loop in the topological space X are both the point $x_0 \in X$, we will say that the loop is *based at* x_0.

For a loop in a topological space X based at x_0, let $[f] = \{g : g$ is a loop in X based at x_0 and $g \underset{p}{\simeq} f\}$. The set $[f]$ is called the **path homotopy equivalence class** of f.

2.3. Theorem. The collection $\{[f] : f$ is a loop based at $x_0 \in X\}$ consists of pairwise disjoint sets.

We will make the collection of all path homotopy equivalence classes of loops based at $x_0 \in X$ into a group. To do this, we need to define an operation on these classes.

Define "multiplication" of these classes by $[f][g] = [fg]$, where $[fg]$ is defined by

$$ fg(t) = \begin{cases} f(2t), & \text{if } 0 \le t \le \frac{1}{2} \\ g(2t - 1), & \text{if } \frac{1}{2} \le t \le 1. \end{cases} $$

Note that fg is the product of paths, and is not composition of functions.

2.4. Exercises.
1) Show that if f and g are loops based at $x_0 \in X$, then fg is also a loop based at x_0.
2) Show that multiplication of path homotopy equivalence classes of loops is well defined, i.e., if f, f', g, g' are loops in X based at x_0 with $f \underset{p}{\cong} f'$ and $g \underset{p}{\cong} g'$, then $fg \underset{p}{\cong} f'g'$. Thus it does not matter which particular loop we use to represent an equivalence class.

For a loop f in X based at x_0, define f^{-1} by $f^{-1}(t) = f(1 - t)$ for $t \in I$. Define e_{x_0} by $e_{x_0}(t) = x_0$ for all $t \in I$.

2.5. Exercises.
1) $[f][f^{-1}] = [f^{-1}][f] = [e_{x_0}]$. To show that $[f][f^{-1}] = [e_{x_0}]$, for example, we have to show that $ff^{-1} \simeq e_{x_0}$. Consider the following picture.

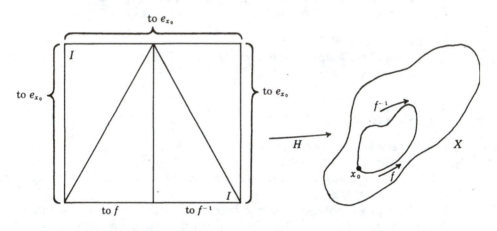

We want $H: I \times I \to X$ such that

$$H(s, 0) = ff^{-1}(s) \quad \text{for all} \quad s \in I$$

$$H(s, 1) = e_{x_0}(s) = x_0 \quad \text{for all} \quad s \in I$$

$$H(0, t) = e_{x_0}(t) = x_0 \quad \text{for all} \quad t \in I$$

$$H(1, t) = e_{x_0}(t) = x_0 \quad \text{for all} \quad t \in I$$

To define H, consider the following picture, in which we indicate where the portion of the edge of the square is to go under H by writing it by that portion.

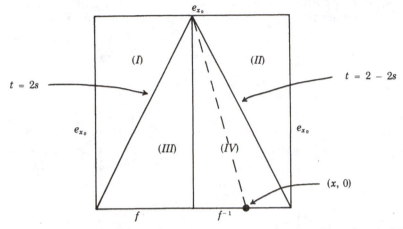

On triangles (*I*) and (*II*), we want *H* to be e_{x_0}. Consider triangle (*III*). We have $t/2 \le s \le \frac{1}{2}$ and $0 \le t \le 1$. Now $t/2 \le s \le \frac{1}{2}$ implies that $t \le 2s \le 1$, so $0 \le 2s - t \le 1 - t$, or $0 \le (2s - t)/(1 - t) \le 1$. Now $[0, 1]$ is the domain of f and it is not hard to show that if we define H on the left half of $I \times I$ by

$$H(s, t) = \begin{cases} x_0, & \text{if } 0 \le s \le t/2 \\ f\left(\dfrac{2s - t}{1 - t}\right), & \text{if } t/2 \le s \le \frac{1}{2}, \end{cases}$$

then *H* is continuous and the part of the square where *H* is *f* shrinks to a point as *t* goes (up) from 0 to 1.

For the right half of the square, we illustrate a different method of finding *H*. Let $(x, 0)$ be a point on the *s*-axis in triangle (*IV*). Draw the line from $(x, 0)$ to $(\frac{1}{2}, 1)$ (this line is shown dotted in the picture). We can write a parametric equation for this line as follows. A point (s, t) in triangle (*IV*) is on this line if and only if there is an a, $0 \le a \le 1$, such that $(s, t) = a(x, 0) + (1 - a)(\frac{1}{2}, 1)$. Then $s = ax + \frac{1}{2}(1 - a)$ and $t = 1 - a$. Solving for *x* gives

$$x = (2s - t)/2(1 - t).$$

Put

$$H(s, t) = \begin{cases} f^{-1}\left(\dfrac{2s - t}{2(1 - t)}\right), & \text{if } \frac{1}{2} \le s \le \dfrac{2 - t}{2} \\ x_0, & \text{if } \dfrac{2 - t}{2} \le s \le 1. \end{cases}$$

Show that $ff^{-1} \underset{p}{\simeq} e_x$.

Show that $[f][f^{-1}] = [f^{-1}][f] = [e_{x_0}]$.

2) For any loop f in X based at x_0, $[f][e_{x_0}] = [e_{x_0}][f] = [f]$. [*Hint*: Consider the following picture for half of this problem.]

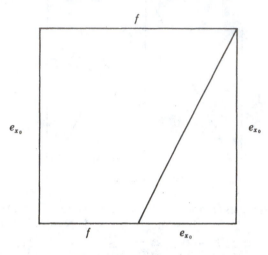

3) For loops f, g, h in X based at x_0, show that $([f][g])([h]) = [f]([g][h])$, i.e., that the multiplication that we have defined is associative. [*Hint*: Consider

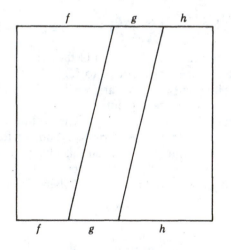

2.6. Theorem. The set of path homotopy equivalence classes of loops based at $x_0 \in X$ is a group under multiplication defined by $[f][g] = [fg]$. This group is denoted by $\pi_1(X, x_0)$ and is called the **fundamental group (first homotopy group, Poincaré group*)** of X at x_0.

* After Jule Henri Poincaré (1854–1912).

2.7. Exercise.

What is the fundamental group of a space consisting of a single point?

Instead of the "fundamental group of X at x_0," it would be nice to have the "fundamental group of X." In other words, we would like to have the fundamental group depend only on the space, and not on the particular point of the space that we base our loops at. There is an important class of spaces in which it turns out that the base point does not matter as far as the structure of the fundamental group is concerned. Recall that a space X is *path connected* if there is a path in X connecting any two given points of X. Recall also that if $(G, *)$ and $(H, **)$ are groups, then a function $\varphi: G \to H$ is an *isomorphism* of G into H if φ is 1-1 and for all $g_1, g_2 \in G$, $\varphi(g_1 * g_2) = \varphi(g_1) ** \varphi(g_2)$. If, in addition, the isomorphism φ is onto, then we say that G and H are *isomorphic*.

2.8. Exercise.

If $\varphi: G \to H$ is an isomorphism of the group G into the group H, then
1) if e is the identity in G, $\varphi(e)$ is the identity in H, and
2) for all $g \in G$, $\varphi(g^{-1}) = (\varphi(g))^{-1}$.

An isomorphism between two groups, then, is an operation preserving function, and isomorphic groups are algebraically the same, just as homeomorphic topological spaces are topologically the same. An isomorphism preserves the algebraic structure; a homeomorphism preserves the topological structure.

2.9. Theorem. Let X be a path connected space. For any two points x_0, $x_1 \in X$, the groups $\pi_1(X, x_0)$ and $\pi_1(X, x_1)$ are isomorphic.

[*Hint for proof:* Consider the following picture. Let p be a path from x_0 to x_1. Define

$$\varphi: \pi_1(X, x_0) \to \pi_1(X, x_1) \quad \text{by} \quad \varphi([f]) = [p^{-1}fp]$$

and show that φ is an isomorphism onto.]

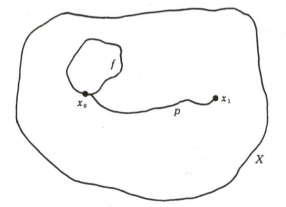

Note that the isomorphism between $\pi_1(X, x_0)$ and $\pi_1(X, x_1)$ defined in the proof of Theorem 2.9 depends on the particular path that we choose to link x_0 and x_1 (actually, it only depends on the path homotopy class of this path). Nevertheless, when X is a path connected space, Theorem 2.9 shows that the fundamental group of X at x_0 does not depend on the particular point x_0 that we choose, so it makes sense to talk simply about the fundamental group of a path connected space. That this is not true in general is the subject of the following exercise.

2.10. Exercise.

Assume that $\pi_1(S^1, x_0)$ is not a trivial group, where S^1 is a circle in the plane and $x_0 \in S^1$ (we will discuss this in the next section). Show that $\pi_1(X, p_0)$ is not the same as $\pi_1(X, x_0)$, where $X = S^1 \cup \{p_0\}$ is the space below.

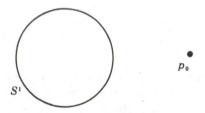

If X and Y are topological spaces and $f: X \to Y$ is continuous, it is natural to ask if there is any relation between the groups $\pi_1(X, x_0)$ and $\pi_1(Y, f(x_0))$. The following theorem answers this question. Recall that if $(G, *)$ and $(H, **)$ are groups, then a function $h: G \to H$ is a *homomorphism* if for all $g_1, g_2 \in G$, $h(g_1*g_2) = h(g_1)**h(g_2)$. (So an isomorphism is a 1-1 homomorphism).

2.11. Exercise.

Like an isomorphism, a homomorphism $h: G \to H$ sends the identity to the identity, and $h(g^{-1}) = (h(g))^{-1}$ for all $g \in G$.

2.12. Theorem. Let X and Y be topological spaces and let $f: X \to Y$ be a continuous function. Then the function $f_*: \pi_1(X, x_0) \to \pi_1(Y, f(x_0))$ defined by

$$f_*([p]) = [f \circ p]$$

is a homomorphism, called the *homomorphism induced by f.*

Furthermore, if f is a homeomorphism of X onto Y, then f_* is an isomorphism of $\pi_1(X, x_0)$ onto $\pi_1(Y, f(x_0))$.

Thus in the particular case of path connected spaces, we can say that homeomorphic spaces have isomorphic fundamental groups. (The path connectivity is necessary only to be able to omit specifying the base point.)

The converse to this statement is false, as you can show in the following exercise.

2.13. Exercise.

The fundamental group of the Euclidean plane with its usual topology is trivial; in other words, it contains only one element.

The process of forming the homomorphism induced by a continuous function also satisfies the following properties.

2.14. Theorem. Let X, Y, and Z be topological spaces and let $f: X \to Y$, $g: Y \to Z$ be continuous functions. Then

1) $(g \circ f)_* = g_* \circ f_*$.
2) $(id_X)_*$ is the identity homomorphism of $\pi_1(X, x_0)$ into itself.

We can use the homomorphism of fundamental groups induced by a continuous function to obtain the following result relating homotopy type to the fundamental group. We will use this result later to get an example of two spaces which are not of the same homotopy type.

2.15. Theorem. Let X and Y be path connected topological spaces which are of the same homotopy type. Then the fundamental groups of X and Y are isomorphic.

3. AN INTUITIVE LOOK AT THE FUNDAMENTAL GROUP OF SOME FAMILIAR SPACES

Computing the fundamental group of a space can be very difficult indeed. In this section, we will look intuitively at the fundamental groups of certain spaces. We do not claim to prove anything here. Instead, like parts of Chapter 3, the material is so interesting that even though some of it is beyond our reach rigorously, we choose to discuss it intuitively and to say "convince yourself" rather than "prove."

3.1. The Circle, S^1.

Let S^1 be a circle in the plane. Choose a point $x_0 \in S^1$. We investigate $\pi_1(S^1, x_0)$.

a) Does it matter what particular point x_0 we choose? In other words, is $\pi_1(S^1, x_0)$ the same as $\pi_1(S^1, x_1)$ for any two points x_0 and x_1 in S^1? A loop in S^1 will be assumed to be based at x_0.

b) Convince yourself that any loop which does not go all the way around the circle is homotopic to the constant loop at x_0 that does not go anywhere. Thus $[e_{x_0}]$ consists of all loops that do not go all the way around the circle.

c) At first thought, it might seem that there is only one non-trivial loop, namely one that starts at x_0, goes around the circle, and stops when it gets back to x_0. Is such a loop that goes clockwise the same as one that goes counterclockwise? Is a clockwise loop related to a counterclockwise loop?

How about a loop that goes around twice? Three times? Once around clockwise and four times around counterclockwise?

Can a loop follow a route like the following?

If so, what is this loop?

d) An *infinite cyclic group on one generator g* is, using multiplicative notation,

$$\{1, g, g^2, g^3, \cdots, g^{-1}, g^{-2}, g^{-3}, \cdots\}$$

and, using additive notation, is

$$\{0, g, 2g, 3g, \cdots, -g, -2g, -3g, \cdots\}.$$

The set of integers under addition is an infinite cyclic group. What is its generator?

e) What is $\pi_1(S^1, x_0)$?

f) Give an example of two spaces that do not have the same homotopy type.

3.2. The Torus.

Let T be the torus

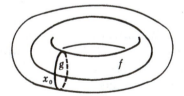

a) Choose $x_0 \in T$. Does the choice of x_0 matter as far as the fundamental group is concerned?

b) Consider two non-trivial loops in T based at x_0: let f go around the hole in the middle and let g go around the cylindrical part. Is $f \underset{p}{\simeq} g$?

c) How many generators does $\pi_1(T, x_0)$ have? What is $\pi_1(T, x_0)$?

d) Convince yourself that the torus is not homeomorphic to a circle.

3.3. Other Examples.

1) What is $\pi_1(X, x_0)$ if X is

a)

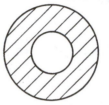

b)

c) S^2, a hollow sphere.

d) $S^2 - \{p\}$, where p is any point on S^2. ($S^2 - \{p\}$ is called a punctured 2-sphere.)

2) Does the fundamental group always "recognize" the number of holes in a space?

3) What is the fundamental group of a simply connected region in the plane?

4. APPLICATIONS

4.1. The Brouwer Fixed Point Theorem. A point $x \in X$ is a *fixed point* of a function $f: X \to X$ if $f(x) = x$. A fixed point of a function is not moved by the function; it remains fixed. Obviously, every constant function of a space into itself has a fixed point, and, just as obviously, not every function of a space into itself has a fixed point. It is an important, interesting, and often difficult problem to decide whether or not a given function of a given space into itself has a fixed point.

In this section, we will prove that every continuous function of the closed unit interval into itself and that every continuous function of the closed unit disk into itself has a fixed point. Both of these results are special cases of a famous theorem of L. E. J. Brouwer which says that if B_n is the closed unit ball in Euclidean n-space (i.e., B_n is the set of all points in E^n whose distance from the origin is no more than 1) then every continuous function of B_n into B_n has a fixed point.

In the case $n = 1$, we can illustrate the theorem with a picture. It should be clear that if $f:I \to I$ is continuous and $f(0) \neq 0$ and $f(1) \neq 1$, then the graph of f will have to cross the line $y = x$. On this line, $f(x) = x$.

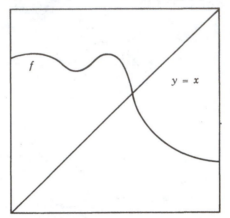

What happens if $f(0) = 0$ or $f(1) = 1$?

Of course this picture is not a proof. To prove the theorem in the case $n = 1$, observe that the subspace $\{0, 1\}$ is not a retract of I (i.e., there is no continuous function of I onto $\{0, 1\}$ which is the identity on $\{0, 1\}$). We will show that if there is a continuous function from I into itself with no fixed point, then $\{0, 1\}$ is a retract of I, which is not true. From this contradiction, we will conclude that every continuous function of I into itself has a fixed point.

4.2. Theorem. Every continuous function of I into I has a fixed point. [*Hint for proof:* Follow the procedure discussed above. Suppose that $f:I \to I$ is continuous with no fixed point. For $x \in I$, define $r:I \to \{0, 1\}$ by $r(x) = 0$ if $f(x) > x$ and $r(x) = 1$ if $f(x) < x$. Using the fact that f is continuous, show that r is a retract of I onto $\{0, 1\}$.

4.3. Exercises
1) How does the proof suggested for Theorem 4.2 break down if it is not assumed that f has no fixed point?
2) Give an example of a function from I into I with no fixed point.
3) Give an example of a continuous function from $(0, 1)$ into itself that has no fixed point.

To prove the fixed point theorem for $n = 2$, we will use the same procedure, and show that if $D = \{(x, y) \in E^2 : x^2 + y^2 \leq 1\}$ and $f:D \to D$ has no fixed point, then there is a retract of D onto S^1, the unit circle. However, in this case it is not so immediately obvious that such a retract cannot exist (unlike $\{0, 1\}$, S^1 is connected). To prove that S^1 is not a retract of D, we can use homotopy theory.*

4.4. Exercise.
Show that if $f: D \to D$ is continuous and has no fixed point, then there is a retract of D onto S^1. [*Hint:* Draw a picture and look at what we did for I.]

4.5. Theorem. S^1 is not a retract of D.
[*Hint:* Consider the fundamental groups of S^1 and D.]

4.6. Theorem. Every continuous function of D into D has a fixed point.

4.7. Exercises.
1) Give an example of a continuous function from

$$\{(x, y) \in E^2 : x^2 + y^2 < 1\}$$

to itself with no fixed point.
2) If you stir water in a glass from below the surface, is it possible for every point on the surface to be in motion at the same time?

4.8. Euclidean spaces. "Obviously," the Euclidean plane is not homeomorphic to Euclidean 3-space. Proving this without the aid of homotopy theory, however, could be very hard indeed. Using homotopy theory makes it easy.

Convince yourself that E^2 and E^3 are not homeomorphic by showing that if they were, then $E^2 - \{(0, 0)\}$ and $E^3 - \{(0, 0, 0)\}$ would also be homeomorphic. Use homotopy theory to convince yourself that this last homeomorphism is impossible.

Is it obvious that $\mathbf{R}$ and E^2 are not homeomorphic?

---

* In the "proof" of this result, assume that we know what the fundamental group of a circle is. We can prove it modulo this fact.

Bibliography

1. Ahlfors, L. V.: *Complex Analysis*, McGraw-Hill, New York, 1953.
2. Bartle, R. G.: *The Elements of Real Analysis*, John Wiley and Sons, New York, 1964.
3. Buck, R. C.: *Advanced Calculus*, McGraw-Hill, New York, 1956.
4. Dugundji, J.: *Topology*, Allyn and Bacon, Boston, 1968.
5. Engelking, R.: *Outline of General Topology*, North Holland, Amsterdam, 1968.
6. Gillman, L., and Jerison, M.: *Rings of Continuous Functions*, Van Nostrand, Princeton, N. J., 1960.
7. Hocking, J., and Young, G.: *Topology*, Addison Wesley, Reading, Mass., 1961.
8. James, G., and James, R. C.: *Mathematics Dictionary*, Van Nostrand Reinhold, New York, 1968.
9. Kelley, J.: *General Topology*, Van Nostrand, Princeton, N. J., 1955.
10. Massey, W.: *Algebraic Topology: An Introduction*, Harcourt, Brace and World, New York, 1967.
11. Nadler, S. B., Jr.: Arc components of certain chainable continua, *Canad. Math. Bull. 14*(2), 1971, 183–189.
12. Rudin, M. E.: A normal space X for which $X \times I$ is not normal, *Bull. A.M.S. 77*, 1971, 246.
13. Sierpinski, W.: *Cardinal and Ordinal Numbers*, Monografie Matematyczne 34, Warsaw, 1965.
14. Simmons, G. F.: *Introduction to Topology and Modern Analysis*, McGraw-Hill, New York, 1963.
15. Singer, I. M., and Thorpe, J. A.: *Lecture Notes on Elementary Topology and Geometry*, Scott Foresman, Glenview, Ill., 1967.
16. Spanier, E. H.: *Algebraic Topology*, McGraw-Hill, New York, 1966.
17. Steen, L. A., and Seebach, J. A., Jr.: *Counterexamples in Topology*, Holt, Rinehart and Winston, New York, 1970.
18. Wilder, R. L.: *Introduction to the Foundations of Mathematics*, John Wiley and Sons, New York, 1965.

Index

A CATALOG OF SELECTED
DOVER BOOKS
IN SCIENCE AND MATHEMATICS

Mathematics-Bestsellers

HANDBOOK OF MATHEMATICAL FUNCTIONS: with Formulas, Graphs, and Mathematical Tables, Edited by Milton Abramowitz and Irene A. Stegun. A classic resource for working with special functions, standard trig, and exponential logarithmic definitions and extensions, it features 29 sets of tables, some to as high as 20 places. 1046pp. 8 x 10 1/2. 0-486-61272-4

ABSTRACT AND CONCRETE CATEGORIES: The Joy of Cats, Jiri Adamek, Horst Herrlich, and George E. Strecker. This up-to-date introductory treatment employs category theory to explore the theory of structures. Its unique approach stresses concrete categories and presents a systematic view of factorization structures. Numerous examples. 1990 edition, updated 2004. 528pp. 6 1/8 x 9 1/4. 0-486-46934-4

MATHEMATICS: Its Content, Methods and Meaning, A. D. Aleksandrov, A. N. Kolmogorov, and M. A. Lavrent'ev. Major survey offers comprehensive, coherent discussions of analytic geometry, algebra, differential equations, calculus of variations, functions of a complex variable, prime numbers, linear and non-Euclidean geometry, topology, functional analysis, more. 1963 edition. 1120pp. 5 3/8 x 8 1/2. 0-486-40916-3

INTRODUCTION TO VECTORS AND TENSORS: Second Edition--Two Volumes Bound as One, Ray M. Bowen and C.-C. Wang. Convenient single-volume compilation of two texts offers both introduction and in-depth survey. Geared toward engineering and science students rather than mathematicians, it focuses on physics and engineering applications. 1976 edition. 560pp. 6 1/2 x 9 1/4. 0-486-46914-X

AN INTRODUCTION TO ORTHOGONAL POLYNOMIALS, Theodore S. Chihara. Concise introduction covers general elementary theory, including the representation theorem and distribution functions, continued fractions and chain sequences, the recurrence formula, special functions, and some specific systems. 1978 edition. 272pp. 5 3/8 x 8 1/2. 0-486-47929-3

ADVANCED MATHEMATICS FOR ENGINEERS AND SCIENTISTS, Paul DuChateau. This primary text and supplemental reference focuses on linear algebra, calculus, and ordinary differential equations. Additional topics include partial differential equations and approximation methods. Includes solved problems. 1992 edition. 400pp. 7 1/2 x 9 1/4. 0-486-47930-7

PARTIAL DIFFERENTIAL EQUATIONS FOR SCIENTISTS AND ENGINEERS, Stanley J. Farlow. Practical text shows how to formulate and solve partial differential equations. Coverage of diffusion-type problems, hyperbolic-type problems, elliptic-type problems, numerical and approximate methods. Solution guide available upon request. 1982 edition. 414pp. 6 1/8 x 9 1/4. 0-486-67620-X

VARIATIONAL PRINCIPLES AND FREE-BOUNDARY PROBLEMS, Avner Friedman. Advanced graduate-level text examines variational methods in partial differential equations and illustrates their applications to free-boundary problems. Features detailed statements of standard theory of elliptic and parabolic operators. 1982 edition. 720pp. 6 1/8 x 9 1/4. 0-486-47853-X

LINEAR ANALYSIS AND REPRESENTATION THEORY, Steven A. Gaal. Unified treatment covers topics from the theory of operators and operator algebras on Hilbert spaces; integration and representation theory for topological groups; and the theory of Lie algebras, Lie groups, and transform groups. 1973 edition. 704pp. 6 1/8 x 9 1/4. 0-486-47851-3

Browse over 9,000 books at www.doverpublications.com

A SURVEY OF INDUSTRIAL MATHEMATICS, Charles R. MacCluer. Students learn how to solve problems they'll encounter in their professional lives with this concise single-volume treatment. It employs MATLAB and other strategies to explore typical industrial problems. 2000 edition. 384pp. 5 3/8 x 8 1/2.   0-486-47702-9

NUMBER SYSTEMS AND THE FOUNDATIONS OF ANALYSIS, Elliott Mendelson. Geared toward undergraduate and beginning graduate students, this study explores natural numbers, integers, rational numbers, real numbers, and complex numbers. Numerous exercises and appendixes supplement the text. 1973 edition. 368pp. 5 3/8 x 8 1/2.   0-486-45792-3

A FIRST LOOK AT NUMERICAL FUNCTIONAL ANALYSIS, W. W. Sawyer. Text by renowned educator shows how problems in numerical analysis lead to concepts of functional analysis. Topics include Banach and Hilbert spaces, contraction mappings, convergence, differentiation and integration, and Euclidean space. 1978 edition. 208pp. 5 3/8 x 8 1/2.   0-486-47882-3

FRACTALS, CHAOS, POWER LAWS: Minutes from an Infinite Paradise, Manfred Schroeder. A fascinating exploration of the connections between chaos theory, physics, biology, and mathematics, this book abounds in award-winning computer graphics, optical illusions, and games that clarify memorable insights into self-similarity. 1992 edition. 448pp. 6 1/8 x 9 1/4.   0-486-47204-3

SET THEORY AND THE CONTINUUM PROBLEM, Raymond M. Smullyan and Melvin Fitting. A lucid, elegant, and complete survey of set theory, this three-part treatment explores axiomatic set theory, the consistency of the continuum hypothesis, and forcing and independence results. 1996 edition. 336pp. 6 x 9.   0-486-47484-4

DYNAMICAL SYSTEMS, Shlomo Sternberg. A pioneer in the field of dynamical systems discusses one-dimensional dynamics, differential equations, random walks, iterated function systems, symbolic dynamics, and Markov chains. Supplementary materials include PowerPoint slides and MATLAB exercises. 2010 edition. 272pp. 6 1/8 x 9 1/4.   0-486-47705-3

ORDINARY DIFFERENTIAL EQUATIONS, Morris Tenenbaum and Harry Pollard. Skillfully organized introductory text examines origin of differential equations, then defines basic terms and outlines general solution of a differential equation. Explores integrating factors; dilution and accretion problems; Laplace Transforms; Newton's Interpolation Formulas, more. 818pp. 5 3/8 x 8 1/2.   0-486-64940-7

MATROID THEORY, D. J. A. Welsh. Text by a noted expert describes standard examples and investigation results, using elementary proofs to develop basic matroid properties before advancing to a more sophisticated treatment. Includes numerous exercises. 1976 edition. 448pp. 5 3/8 x 8 1/2.   0-486-47439-9

THE CONCEPT OF A RIEMANN SURFACE, Hermann Weyl. This classic on the general history of functions combines function theory and geometry, forming the basis of the modern approach to analysis, geometry, and topology. 1955 edition. 208pp. 5 3/8 x 8 1/2.   0-486-47004-0

THE LAPLACE TRANSFORM, David Vernon Widder. This volume focuses on the Laplace and Stieltjes transforms, offering a highly theoretical treatment. Topics include fundamental formulas, the moment problem, monotonic functions, and Tauberian theorems. 1941 edition. 416pp. 5 3/8 x 8 1/2.   0-486-47755-X

Browse over 9,000 books at www.doverpublications.com

Mathematics–Algebra and Calculus

VECTOR CALCULUS, Peter Baxandall and Hans Liebeck. This introductory text offers a rigorous, comprehensive treatment. Classical theorems of vector calculus are amply illustrated with figures, worked examples, physical applications, and exercises with hints and answers. 1986 edition. 560pp. 5 3/8 x 8 1/2.	0-486-46620-5

ADVANCED CALCULUS: An Introduction to Classical Analysis, Louis Brand. A course in analysis that focuses on the functions of a real variable, this text introduces the basic concepts in their simplest setting and illustrates its teachings with numerous examples, theorems, and proofs. 1955 edition. 592pp. 5 3/8 x 8 1/2.	0-486-44548-8

ADVANCED CALCULUS, Avner Friedman. Intended for students who have already completed a one-year course in elementary calculus, this two-part treatment advances from functions of one variable to those of several variables. Solutions. 1971 edition. 432pp. 5 3/8 x 8 1/2.	0-486-45795-8

METHODS OF MATHEMATICS APPLIED TO CALCULUS, PROBABILITY, AND STATISTICS, Richard W. Hamming. This 4-part treatment begins with algebra and analytic geometry and proceeds to an exploration of the calculus of algebraic functions and transcendental functions and applications. 1985 edition. Includes 310 figures and 18 tables. 880pp. 6 1/2 x 9 1/4.	0-486-43945-3

BASIC ALGEBRA I: Second Edition, Nathan Jacobson. A classic text and standard reference for a generation, this volume covers all undergraduate algebra topics, including groups, rings, modules, Galois theory, polynomials, linear algebra, and associative algebra. 1985 edition. 528pp. 6 1/8 x 9 1/4.	0-486-47189-6

BASIC ALGEBRA II: Second Edition, Nathan Jacobson. This classic text and standard reference comprises all subjects of a first-year graduate-level course, including in-depth coverage of groups and polynomials and extensive use of categories and functors. 1989 edition. 704pp. 6 1/8 x 9 1/4.	0-486-47187-X

CALCULUS: An Intuitive and Physical Approach (Second Edition), Morris Kline. Application-oriented introduction relates the subject as closely as possible to science with explorations of the derivative; differentiation and integration of the powers of x; theorems on differentiation, antidifferentiation; the chain rule; trigonometric functions; more. Examples. 1967 edition. 960pp. 6 1/2 x 9 1/4.	0-486-40453-6

ABSTRACT ALGEBRA AND SOLUTION BY RADICALS, John E. Maxfield and Margaret W. Maxfield. Accessible advanced undergraduate-level text starts with groups, rings, fields, and polynomials and advances to Galois theory, radicals and roots of unity, and solution by radicals. Numerous examples, illustrations, exercises, appendixes. 1971 edition. 224pp. 6 1/8 x 9 1/4.	0-486-47723-1

AN INTRODUCTION TO THE THEORY OF LINEAR SPACES, Georgi E. Shilov. Translated by Richard A. Silverman. Introductory treatment offers a clear exposition of algebra, geometry, and analysis as parts of an integrated whole rather than separate subjects. Numerous examples illustrate many different fields, and problems include hints or answers. 1961 edition. 320pp. 5 3/8 x 8 1/2.	0-486-63070-6

LINEAR ALGEBRA, Georgi E. Shilov. Covers determinants, linear spaces, systems of linear equations, linear functions of a vector argument, coordinate transformations, the canonical form of the matrix of a linear operator, bilinear and quadratic forms, and more. 387pp. 5 3/8 x 8 1/2.	0-486-63518-X

Browse over 9,000 books at www.doverpublications.com

Mathematics–Geometry and Topology

PROBLEMS AND SOLUTIONS IN EUCLIDEAN GEOMETRY, M. N. Aref and William Wernick. Based on classical principles, this book is intended for a second course in Euclidean geometry and can be used as a refresher. More than 200 problems include hints and solutions. 1968 edition. 272pp. 5 3/8 x 8 1/2. 0-486-47720-7

TOPOLOGY OF 3-MANIFOLDS AND RELATED TOPICS, Edited by M. K. Fort, Jr. With a New Introduction by Daniel Silver. Summaries and full reports from a 1961 conference discuss decompositions and subsets of 3-space; n-manifolds; knot theory; the Poincaré conjecture; and periodic maps and isotopies. Familiarity with algebraic topology required. 1962 edition. 272pp. 6 1/8 x 9 1/4. 0-486-47753-3

POINT SET TOPOLOGY, Steven A. Gaal. Suitable for a complete course in topology, this text also functions as a self-contained treatment for independent study. Additional enrichment materials make it equally valuable as a reference. 1964 edition. 336pp. 5 3/8 x 8 1/2. 0-486-47222-1

INVITATION TO GEOMETRY, Z. A. Melzak. Intended for students of many different backgrounds with only a modest knowledge of mathematics, this text features self-contained chapters that can be adapted to several types of geometry courses. 1983 edition. 240pp. 5 3/8 x 8 1/2. 0-486-46626-4

TOPOLOGY AND GEOMETRY FOR PHYSICISTS, Charles Nash and Siddhartha Sen. Written by physicists for physics students, this text assumes no detailed background in topology or geometry. Topics include differential forms, homotopy, homology, cohomology, fiber bundles, connection and covariant derivatives, and Morse theory. 1983 edition. 320pp. 5 3/8 x 8 1/2. 0-486-47852-1

BEYOND GEOMETRY: Classic Papers from Riemann to Einstein, Edited with an Introduction and Notes by Peter Pesic. This is the only English-language collection of these 8 accessible essays. They trace seminal ideas about the foundations of geometry that led to Einstein's general theory of relativity. 224pp. 6 1/8 x 9 1/4. 0-486-45350-2

GEOMETRY FROM EUCLID TO KNOTS, Saul Stahl. This text provides a historical perspective on plane geometry and covers non-neutral Euclidean geometry, circles and regular polygons, projective geometry, symmetries, inversions, informal topology, and more. Includes 1,000 practice problems. Solutions available. 2003 edition. 480pp. 6 1/8 x 9 1/4. 0-486-47459-3

TOPOLOGICAL VECTOR SPACES, DISTRIBUTIONS AND KERNELS, François Trèves. Extending beyond the boundaries of Hilbert and Banach space theory, this text focuses on key aspects of functional analysis, particularly in regard to solving partial differential equations. 1967 edition. 592pp. 5 3/8 x 8 1/2.
0-486-45352-9

INTRODUCTION TO PROJECTIVE GEOMETRY, C. R. Wylie, Jr. This introductory volume offers strong reinforcement for its teachings, with detailed examples and numerous theorems, proofs, and exercises, plus complete answers to all odd-numbered end-of-chapter problems. 1970 edition. 576pp. 6 1/8 x 9 1/4. 0-486-46895-X

FOUNDATIONS OF GEOMETRY, C. R. Wylie, Jr. Geared toward students preparing to teach high school mathematics, this text explores the principles of Euclidean and non-Euclidean geometry and covers both generalities and specifics of the axiomatic method. 1964 edition. 352pp. 6 x 9. 0-486-47214-0

Browse over 9,000 books at www.doverpublications.com